To Amelia and Everly,

The wind will blow, and the world will change.

May you always find your own ground,

and know the Light that holds you up.

Standing in the Wind

Seven Companies Reshaping an Age

Copyright Page

Standing in the Wind

Author: Enze Yang (pen name)
This work is published under a pen name. The author's legal identity is on file with the publisher.

This is a work of nonfiction. The content is based on publicly available information, the author's analysis, and personal interpretation, and may include literary expression and subjective viewpoints. All opinions expressed in this book are solely those of the author and do not represent the official views of any companies mentioned.

Portrayals of individuals, including interpretations of their decisions and motivations, are based on publicly available information and reflect the author's independent analysis. They are not authorized or endorsed by the individuals described.

ISBN: 979-8-9944390-4-3

Cover design: Mark K
Publisher: Enze Yang Press LLC

First Edition · 2026
Printed in the United States of America

Author's Note

I must begin with a confession:
I am not someone trained in technology.

In the face of vast computational landscapes
and intricate algorithmic logic,
I remain, at heart, an outsider.

The descriptions and reflections in this book
are not rigorous technical analyses,
nor are they authoritative conclusions
about the inner worlds of those involved.

They are simply the observations and impressions
of someone walking on lower ground,
as the great winds of this age pass through.

I do not know where technology will ultimately lead.
Perhaps even before this book is finished,
it will have already changed direction once again.

Technology does not always move along a stable path.
At times, it feels more like a force
in constant reconfiguration.

Before old structures have fully settled,
new storms have already arrived.

Forms are shifting.
Entrances are moving.
Platforms are being replaced.
Computing power continues to rise.

Yet beneath all this turbulence,
I have gradually come to see
that some things do not change.

The open-source movement of Meta
was the first moment I realized
that technology is not only evolving—
it is also dissolving existing orders.

Boundaries that once appeared solid
may lose their meaning in an instant.

But what truly holds up a company,
what sustains an era,
has never been a single technology.

It is the culture and convictions
hidden deep within.

Products will disappear.
Platforms will be replaced.
Forms will be reconstructed.

But culture
often outlives them all.

It determines how a company stands in the wind—
what it fears,
what it refuses,
and what it insists on.

As I write about structure, order, and power,
I am aware that what these words describe
is not a completed world,
but an age still in the process of becoming.

In the earlier chapters of this book,
I attempt to understand
the foundations of companies—
their missions, cultures,
and the temperaments of their leaders.

These are what sustain them at their core.

But companies do not remain still.
The winds of the age
reshape their forms, strategies, and positions.

And so, in the concluding chapters,
my gaze turns toward change itself.

Essence and form do not contradict each other.
Foundations allow a company to stand;
the wind reveals its structure.

Stillness contains movement,
and movement carries stillness.

This book unfolds
in the space between the two.

Perhaps I have misread certain technical details.
Yet it is precisely this distance
that has allowed me to step away
from the noise of data,

and try to discern,
behind the cold surfaces of silicon,
the human will and courage
still burning within.

If these words touch upon something true,
it is not because my insight is especially deep,
but because—

those who stand in the wind
are already giving off light.

In the world of artificial intelligence,
there is another company

equally worthy of being written about:
Anthropic.

But when I began this book,
technology was still unfamiliar to me.

Some companies entered my field of vision;
others passed me by.

Anthropic, perhaps, belongs to the latter.

Perhaps one day in the future,
when I look back on this era again,
I will realize
that many stories escaped my sight at the time.

The age is still unfolding.
I have only seen
a part of it.

Prologue · Standing in the Wind

After retirement,
there was no longer a need to rush for a living.
Life suddenly slowed.

On Sundays, I go to church.
On weekdays, I read the Bible at home.
There is wind outside the window,
and light within the room.

The days are quiet,
and gentle.

At this stage of life,
one begins to understand something—

God has not laid out every step for us.
He has placed the power of choice into our hands.

Perhaps that is the greatest freedom He gives,
and also the greatest respect.

We may turn away,
or we may turn back and follow.

As for me,
after walking through most of my life,
I have only just learned one simple thing—

to choose to be connected with Him,
and to let that choice be my response.

Recently, I have been reading *Judges*, *Ruth*, and the
Gospel of *Luke*.
My heart has felt like parched grain meeting rain,
slowly, gently, being nourished.

When I read *Judges*, I saw myself.

At fifteen,
in a remote and impoverished mountain village,
entering high school from middle school
was an almost unreachable dream.

Hope was nearly nonexistent.

For the first time,
I spoke—
and prayed to a God I did not yet know.

I asked for His protection,
for His care.

He showed me mercy.

And yet, filled with joy,
I turned away,
left Him behind,
and threw myself into books and the future.

—That, too, was a choice.

Twenty years later, I came to the United States.
I studied, worked, and struggled through life.

Until one day, while reading *Genesis*,
God suddenly opened my eyes—and my heart.

For the first time,
I truly knew who He is.

I accepted the Lord Jesus.

—That, too, was a choice.

Now, another thirty-five years have passed.
The young girl has become an old woman.

Reading *Judges* again,
I realize that those pages
have been writing my life all along—

calling out,
being delivered;
failing,
calling out again;
being delivered again.

A cycle, repeating itself.

I have wept, yet not repented;
sought grace, yet not turned back to Him.
Turning in circles, again and again.

And yet—

even so,
God has never abandoned me.

He has remained by my side,
silent, yet guarding.

Ruth, however, showed me another path.

A Moabite woman,
without miracles, without spectacle—
she simply chose to believe,
and then walked forward, step by step.

She did not wait for conditions to improve.
She did not wait for the future to become clear.

She made a decision—
to stay.

That was not impulse.
It was a choice.

Faith kept her from turning back.
Love carried her to the end.

God's salvation,
it turns out,
is not always dramatic.

It is quietly accomplished
within the ordinary.

Not because she was better,
but because—
she was willing.

Willing to obey,
willing to pay the cost,
willing, even when the outcome could not be seen,
to continue choosing to believe.

Judges shows me the chaos of man.
Ruth shows me the restoration of God.

When people do as they please, the world unravels.
When people live by covenant, history continues.

God is not waiting for a group of people
who suddenly become better.
He is searching for a few—

who consistently choose what is right.

In the darkness,
they do not seize,

they do not cross the line,
they do not give up.

Thread by thread,
they stitch redemption back into history.

Sometimes I ask myself:
What use am I, a person like this?

Ruth answers me—

one person willing to obey
is enough
for God to begin again.

Only later did I understand:

Calling out may bring salvation,
but it does not bring standing.

To stand,
one must be connected.

Not by willpower,
not by personality,
but—

by abiding in the Lord.

The wind will continue to blow.
But those who are connected to Him
can stand in the wind.

And so I cannot help but wonder:

If this is true for people,
what about companies?

What allows a company
to stand in the wind?

Scale?
Technology?
Market position?

Slowly, I realized—

none of these.

I have seen too many companies
flourish in times of ease,
only to collapse in adversity.

And I have seen others
bend without breaking in the storm,
and rise again from the valley.

The difference between them
does not lie in scale,
nor in technology,
nor in market share.

It lies deeper—

in mission.

Mission is the reason a company exists.

It is not a slogan on a wall,
but the direction a CEO still holds
under the greatest pressure.

It tells the entire organization:
who we are,
why we are here,
and what we must never become.

A company without mission
is like a tree without roots.
When the wind blows,
it does not even know which way to fall.

Yet mission itself
must be carried by someone.

That person
is the CEO.

A CEO is not merely a decision-maker.
He is the guardian of the company's soul.

Through his choices,
again and again,
he tells the organization—

what we believe,
how we treat one another,
how we stand in the dark.

Over time,
those choices
become culture.

Culture is not policy.
Not process.
Not a training manual.

Culture is like air—

it determines how people decide
when no one is watching.

The CEO shapes the culture,
and the culture shapes the CEO.

They reinforce one another,
influence one another,
and deepen one another.

A soft-hearted CEO
will gradually nurture an organization that listens.

A CEO who trusts people
will grow a team that trusts one another.

And the reverse is also true—

when culture begins to decay,
if the CEO does not hold the line,
he will be consumed by it.

This is not management theory.
This is the ecology of the soul.

Within this ecology,
there is also another group—

the team that chooses to follow.

Without them,
mission is only a monologue.
Without them,
culture is only words on a wall.

A company's true ability to stand
has never been the story of a single hero.

It is the story of a group of people
moving in the same direction,
step by step,
together.

Mission gives direction.
Culture gives strength.
The CEO guards the soul.
The team brings everything into reality.

When these four are connected,
a company can stand in the wind.

Not because it is strong enough,
but because it is connected.

Like branches abiding in the vine.
Like Ruth choosing to stay.
Like what Satya Nadella learned
by the bedside—

true standing
has never come from oneself.

It comes from connection.

The wind is still blowing.

Some companies bend.
Some fall.

And some—

stand quietly,
unmoved.

I want to write their stories.

Not because they are perfect,
but because—

they have chosen connection.

Their stories
begin on the next page.

Part 1 · The Manifestation of Fire— OpenAI

Chapter 1 · Fire

From Nonexistence to the Center of the World — Final Polished Version

For a long time,
intelligence existed only within the human mind.

It lived in the mind,
in language,
in experiences that could not be replicated.

Machines could compute.
They could execute.
They could repeat.

But they could not understand.

They had no real answers.
They followed instructions,
but could not participate in thought itself.

Humanity stood at the center of intelligence.
This was one of the deepest assumptions of civilization.

Until one day,
that assumption began to shift.

It was a subtle change—
yet it altered everything.

For the first time,
intelligence no longer belonged solely to the human
mind.
It began to exist within another structure.

OpenAI was one of the starting points of this shift.

At the beginning,
it was not a typical company.

It had no mature business model,
no stable revenue,
not even a clear path.

What it had
was a goal that seemed almost unreal:

to create artificial general intelligence.

Not a better product.
Not faster computation.

But intelligence itself.

Few truly understood this goal.
Because what it sought to create
was not a product—
but a new form of existence.

Then, the fire began to spread.

GPT-2 was only a probe.
GPT-3 made the world realize that something had begun.

And when ChatGPT appeared,
the fire was seen by everyone.

But what changed
was not merely visibility.

Before ChatGPT,
artificial intelligence still lived in the laboratory.

It lived in papers,
in models,
in weights trained over time.

People knew it was becoming more powerful—
but they could not truly touch it.

Until ChatGPT.

For the first time,
an ordinary person needed no code,
no interface,
no understanding of models.

Only a sentence—
and intelligence responded.

Not search results.
Not links.
Not arranged information.

But language.

A complete sentence.
A response.
A conversation that could continue.

In that moment,
something quiet—yet irreversible—happened:

for the first time,
humanity began to speak with intelligence itself.

It was no longer just a tool.
It became something to engage with.

Humans no longer only used technology—
they began to converse with it.

From that moment on,
AI was no longer just an increase in capability,
but a change in relationship.

It is still young.
Still unstable.
Still in need of fuel.
Still dependent on structure.

But the air
has already changed.

Chapter 2 · The Idealist

The One Who Believed Fire Could Exist — Final Polished Version

In human history,
most of those who changed the world were engineers.

They built machines.
They built systems.
They built structures that could be touched.

But there are a few—very few—
who do not build directly.

They simply see
a possibility that does not yet exist.

And then,
they allow the world
to begin moving in that direction.

Sam Altman
is one of them.

He is not the greatest programmer.
Nor the greatest scientist.

He did not invent new algorithms.
He did not design new chips.

What he does
is something else.

He believes
that intelligence can be created.

He did not invent the fire.
He made space for it.

Not as a tool—
but as a form of being.

At the beginning of OpenAI,
this goal seemed almost unreal.

To create artificial general intelligence
was not a business plan—
but a hypothesis.

There was no clear path.
No guaranteed outcome.
Not even a defined timeline.

It was only
a direction.

Most people would not choose such a goal.

But Altman did.

Not because it was easy—
but because it mattered.

He understood
that fire cannot exist alone.

It needs fuel.
It needs structure.
It needs the world
to make room for it.

So he began to bring others in.

Microsoft became the foundation.
NVIDIA became the fuel.
Developers became the carriers.

This was not an isolated creation.
It was the formation of a structure.

What makes Altman different
is not what he built—
but what he set in motion.

He made the world
take seriously
something not yet fully born.

He made investors believe.
He drew in engineers.
He moved the industry
toward this idea.

He did not create the fire.
But he made it possible.

And once something becomes possible—
the world can no longer pretend
it was never there.

Because of OpenAI,

the world began to change.
But the change did not remain on the surface.
At a deeper level,
the structure itself began to shift.
It is not one change among many,
but the first to touch the structure.

Chapter 3 · Between Mission and Survival

From the very beginning,
OpenAI was different.

It was not built for the market.
Not for share.
Not for stock price.

Its charter states that:
when mission and profit conflict—
mission comes first.

This is an unusual choice.

But OpenAI does not exist outside the real world.

It speaks of mission.
And it speaks of compute costs.

It speaks of risk.
And it speaks of funding cycles.

Training models requires immense compute.
Compute requires chips.
Chips require capital.
And capital requires a path to return.

So it designed a rare structure:

a nonprofit at the top,
and a capped-profit company
to operate the business.

Idealism and the market
are bound to the same vessel.

This is not pure idealism.
Nor is it pure capital logic.

It is a balance.

But balance is never still.

As technology approaches the edge of the unknown,
the question is no longer who is more intelligent—

but who decides.

This structure is answering a deeper question:

when mission and survival
pull in different directions—
which one comes first?

And the market's response
keeps pulling reality back to the table.

In that tension,
OpenAI reveals its most truthful form:

it cannot detach from capital,
and it cannot abandon its mission.

It must survive—
and it must hold a line

that cannot be seen.

That line does not appear
in financial statements,
nor in stock prices.

It exists in a deeper unease:

when intelligence begins to approach
the boundary of the human—
are we still present?

OpenAI does not reject expansion.
It does not refuse profit.

It stands, instead,
between fire and the ledger—
asking—again and again:

are we still aligned?

There is no final answer.

But the question
must continue to be asked.

Chapter 4 · The One Cast Out by Fire

**When Systems Attempt to Control the Future —
Final Polished Version**

From its earliest days,
OpenAI understood
that what it sought to ignite
was not an ordinary fire.

Its mission
was written before profit.

Its board
did not fully represent capital.

Its purpose
was not to maximize returns—
but to ensure
that the birth of intelligence
would not slip beyond human control.

This was a system
designed to place boundaries
around fire.

In November 2023,
Sam Altman
was suddenly removed by the board.

There was no warning.
No clear or sufficient public explanation.
No visible transition.

The decision arrived
like a silent fracture.

For the first time,
the world saw
that inside OpenAI,
things were not as stable
as they appeared.

This did not feel like a decision
born of failure.

It felt like a divergence
emerging at the speed of success.

The fire
had become real enough—
powerful enough—
to make the system

itself uneasy.

What followed
changed something deeper.

Employees began to take a position.
Engineers publicly supported Altman.

Microsoft moved quickly
to signal
that he still had a place.

This was not a simple change of role.
It was a decision about direction.

Not the system—
but people
determined
where the fire would go.

Days later,
Altman returned.

The board was restructured.
The system regained its stability.

It seemed
as if everything
had returned
to its original place.

But only a few recognized—
something deeper
had already changed.

Systems can define boundaries.

But once people make a choice—
boundaries
no longer belong
to the system alone.

Chapter 5 · Fire Belongs to No One

**When Intelligence Becomes Part of the World —
Final Polished Version**

Fire
has never truly belonged to anyone.

Humans can ignite it,
guide it,
hold it for a time.

But its nature
is not ownership.

Its nature
is to burn.

As long as the conditions exist,
it continues.
As long as there is fuel,
it spreads.

It does not belong
to any one person,
nor to any single structure.

Artificial intelligence
is becoming such a presence.

OpenAI lit the fire—
and for the first time,
the world saw
intelligence
beyond the human mind.

It became the center.
The beginning.
The first visible light.

But fire
does not remain
at its origin.

Meta allowed it to spread.
NVIDIA provided the fuel.

And so,
intelligence no longer belongs
to a single company.

It is becoming
part of the structure of the world.

OpenAI still stands there—
as one of the places
where the fire was first seen.

It continues to push it forward,
to carry it further.

But it is no longer
the only vessel.

The fire
has entered the air.

This is not the end.

This is the beginning.

No one knows
where it will lead.

But one thing
can no longer be changed:

the fire
is now in the world.

And humanity
will walk with it.

Epilogue · After the Fire

I sat for a long time
after writing this.

The wind
was still moving softly.

People on the street
went on as they always had—
unaware
that a new era
was already taking shape.

Those companies
are still there.

Those machines
still run.

Those chips
continue to burn in silence.

They do not speak.
They do not explain.

They simply continue—
like forces deep beneath the earth,
slow,
steady,
unseen.

Humanity has lit the fire.

But it does not yet understand
where it will lead.

Perhaps
no one truly knows.

Perhaps
the answer
belongs to no company—
and to no one

at all.

Part 2 · The Rebuilder of Structure — Microsoft

On the morning of February 4, 2014,
Satya Nadella walked into Microsoft's headquarters.

It was a winter morning.
In the air, there was a tension not yet named,
and a quiet sense of expectation.

This great vessel
had been sailing for decades.
It had known brilliance—
and it had known stagnation.

Fiefdoms had taken root.
Processes had grown heavy.
The flame of innovation
had been slowly smothered,
layer by layer,
by the instinct of self-preservation.

It was at such a moment
that he succeeded
Steve Ballmer,
becoming the company's third CEO.

He was not an outsider
brought in to impose change.

By then,
he had already spent twenty-two years inside Microsoft—
moving through nearly every core division,
learning its structure,
and its wounds.

Chapter 1 · Healing

Rebuilding Microsoft's Culture

Microsoft's revival did not begin with technology.
Its true beginning
was a quiet act of healing.

At that time, Microsoft did not lack technology.
Nor did it lack market presence.
What it lacked
was relationship.

It lacked trust.
It lacked connection.
It even lacked the kind of atmosphere
that allows people to draw near to one another
and move forward together.

The company was like a giant—
strong in body,
yet weary in soul.

Walls stood between divisions.
Engineers guarded themselves against one another.

Teams fought for credit,
competed for resources,
and undermined each other.

Everyone was intelligent.
And yet the company as a whole
felt stranded—
like islands, cut off from one another.

What Nadella saw
was not a problem of technology.
It was a problem of emotion.
Of relationships.
Of culture.

So the first thing he did
was not to set a strategy.
It was to heal.

He did not raise slogans.
He did not rule with force.

The first thing he did
was not to launch a new product—
but to write a letter.
A letter to everyone.

In that letter,
he spoke of mission,
of learning,
of empathy.

In a Silicon Valley
that prizes strength and logic,
this felt almost out of place.

His tone was gentle,
yet clear.

He did not criticize the past.
Nor did he exaggerate the future.
He asked only two simple questions:

— Who are we?
— Why do we exist?

He said:
Microsoft should not be
a "know-it-all" company,
but a "learn-it-all" company.

The core of technology
is not code—
but the ability
to understand the needs of others.

That letter was like a light wind.
No slogans—
and yet, for the first time,
people sensed:
Microsoft was about to change.

Not from products,
but from the soul.

Not from competition,
but from humility.

That letter
was like the first needle to fall—
so light it was almost unheard,
yet it quietly shifted
the direction of the entire ship.

Later,
he asked all senior leaders
to read *Nonviolent Communication*.

People mocked it
as something like "kindergarten education."

But he understood—
language shapes culture.

Those who were used to pounding the table
and raising their voices
found themselves, for the first time,
awkwardly saying in meetings:

"I feel…"

The room fell still.

And in that moment,
a kind of silence—long absent—returned.

He did not repair the company with code.
He tried to repair it with language.

An organization
that knows only how to speak of "winning" and "losing"
will, in the end,
lose to its time.

What he handed them
was not just a book,
but a mirror—
one that allowed a group of people
accustomed to fighting
to see, for the first time,
the lack within themselves.

Microsoft began to change—
from "winning over others"
to "understanding others";
from "proving oneself"
to "connecting with others."

He dismantled the silos.
He allowed teams to speak again.
He allowed engineers to trust engineers again.
He allowed products to understand users again.
He allowed management to understand people again.

Slowly, Microsoft changed—
from an empire of internal competition
into a community of connection.

This was not a management technique.
It was the rebuilding of the soul.

Nadella understood:

A company divided within
cannot embrace the world.
A company without trust
cannot truly be open.
A company that has not been healed
cannot stand.

And this
is where Microsoft's revival began.

Not with technological breakthroughs,
but with softened hearts,
with fallen walls,
with people
learning to see one another again.

Healing
was the first step
for it to stand again.

And it was
the most essential one.

Chapter 2 · Openness

From a Closed Empire to Embracing the World

After healing,
what came to Microsoft
was not another technological revolution,
but a change both harder
and deeper—
a turning of posture.

In its earlier era,
Microsoft was like a towering castle.
Its walls were thick.
Its gates were narrow.

Inside, people competed with one another.
Outside, others were kept at a distance.

It had once been powerful—
and in that power,
it had grown closed, proud, and isolated.

Power, sustained for too long,
can make one forget
that the world is still changing.

After Satya Nadella took over,
the second thing he did
was not to launch a new product,
but—
to open the gates.

Not to push them slightly ajar,
but to remove them entirely.
To let the wind in.
And to let people in.

He led Microsoft
to embrace Linux—
something once regarded as an outsider.

At the time,
this was almost unthinkable.

A company that had once treated Linux as a threat—
even as an enemy—
now extended its hand and said:
"We can move forward together."

In that moment,
it was not technology that yielded—
it was pride.

He led Microsoft
to embrace iOS,
and Android as well.

No longer insisting
that "only Windows is the future,"
but recognizing that:
the world is diverse,
users have choices,
and ecosystems follow their own paths.

Microsoft no longer asked the world
to enter its garden.
It chose instead
to step into the world.

After opening itself,
Microsoft moved one step further—
into open source.

Open source may appear
to be a technical term.
But what it truly changes
is not code—
it is relationships.

It changes how software is built.
And how people collaborate.

Source code becomes visible—
to be read, modified, used, and shared by all.

It sounds simple.
But it requires letting go of control,
letting go of exclusivity,
even letting go of that deep-rooted fear
of losing control.

Open source is a posture:
not "I give you code,"
but "we build this together."

Open source is a culture:
transparency,
sharing,
peer review,
collective improvement.

Open source is also a force—
Kubernetes reshaped cloud computing.
GitHub brought developers from around the world
to the same table.

Software no longer belongs
to a single company—
it becomes the work
of a global community.

Microsoft once stood
against all of this.

Now,
it chose to stand among the people.

Not only using open source,
but contributing to it,
maintaining it,
investing in it—
moving from an opponent
to one of its largest contributors.

It even acquired GitHub—
not to possess it,
but to protect
a home for developers.

Behind all of this
was a new disposition:
openness,
inclusiveness,
humility.

Nadella understood:

A company that believes only in itself
will grow smaller.
A company willing to trust the world
will grow larger.

Openness is not a strategy.
Not public relations.
Not even calculation.

Openness
is vitality.

And so Microsoft changed—
from a closed empire
to a companion.

From "I possess everything"
to "I am willing to work with you."
From "I must win"
to—
"we win together."

In that moment,
Microsoft truly stood again.

Not because its technology was stronger—
but because—
its heart had softened,
its gates had opened,
and its path had widened.

Openness
allowed it to breathe again.
Openness
allowed the world to receive it again.
Openness
allowed it to have a future again.

Chapter 3 · Alignment with OpenAI

The Sky Opens

In December 2015,
OpenAI announced its founding.

Silicon Valley stirred.

It was a gathering
of some of the world's leading scientists,
entrepreneurs,
and idealists.

They declared that their goal
was not profit,
but—
to ensure that artificial intelligence
benefits all of humanity.

Soon,
the list of founding supporters circulated.

Amazon was there.
Tech giants were there.
Pioneers of the industry were there.

Only—
Microsoft was not.

This was no oversight.
It was the world's quiet judgment:

Microsoft
no longer represented the future.

At the time,
this was not surprising.

Although Microsoft had begun to change
under Satya Nadella,
that change was still limited.

It had only just stepped out
of its closed past.
Its healing had only begun.
Its openness
had not yet been seen.

It was still
outside the door.

When this news reached Microsoft,
Nadella did something small—
yet often retold.

He sent a brief email to his team:

"Why are we not on the list?"

There was no anger.
No resentment.

Only a rare clarity.

Not,
"Why were we not invited?"
but—
"Why are we not yet worthy of being invited?"

In that moment,
Microsoft realized for the first time:

the table of the future
had already moved—
and it was not seated at it.

That email
became the true beginning of the story.

It was like a beam of light,
revealing a direction
Microsoft had not yet reached.

What Nadella saw
was not absence itself,
but something deeper—

the entry point of a new era
was quietly opening elsewhere.

If Microsoft wanted to return
to the center,
it would have to learn something
it had never truly learned:

to walk with the future—
not to claim ownership of it.

And so,
before the company was fully ready,
before its culture was fully rebuilt,
before the world had fully accepted it,
Nadella chose to move toward OpenAI.

Not to chase a trend.
Not to seize a moment.

But because he sensed—
the future of computing
would no longer belong to devices,
but to intelligence itself.

The collaboration began quietly.

At the time,
almost no one understood
what it meant.

It was not an acquisition.
Not a massive investment.
It drew little media attention.

It was simply a decision:

OpenAI would run on Azure.

To outsiders,
this looked like an ordinary cloud contract.

But in Nadella's eyes,
it was the first thread
leading Microsoft back into the future.

OpenAI needed compute.
Microsoft needed an entry.

Two rivers,
in the dark,
heard each other for the first time.

At that time,
OpenAI was still small.

No product.
No business model.

Only a group of persistent researchers,
asking an almost naïve question:

"Can intelligence truly understand the world?"

And Microsoft, too,
was still in recovery.

It needed a spark.

So two forces,
in a corner unnoticed,
quietly bound themselves together.

Like a seed in the soil—
silent,
small,
yet already
reshaping destiny.

Three years later,
OpenAI shifted
to a capped-profit structure.

It needed a long-term,
stable,
trusted partner.

Microsoft committed one billion dollars.

Doubts followed.

Why invest in a company with no product?
Why not build it internally?
Why place such a bet
on a research organization?

But Nadella was never looking
at the present.

He was looking
ten years ahead.

Microsoft might build faster models—
but it could not create
a team more free,
more focused,
more willing to push boundaries.

Technology can be replicated.
Spirit cannot.

OpenAI had the spirit Microsoft lacked.
Microsoft had the systems,
the engineering discipline,
and the global infrastructure OpenAI lacked.

This was not an acquisition.
Not control.

It was—
alignment.

A mutual completion.

When GPT-3 emerged,
Microsoft realized:
this was not just a model—
it was an entry.

When ChatGPT arrived,
the world realized:
Microsoft had placed its bet
on the future.

When Copilot emerged,
Microsoft itself realized:

this was not merely
a technological shift—
it was a rewriting of its destiny.

By 2023,
the partnership between Microsoft and OpenAI
had deepened into full alignment.

Microsoft committed tens of billions of dollars.
Azure became the core infrastructure
for OpenAI.

OpenAI's models
became the intelligent foundation
across Microsoft's product ecosystem.

In that moment,
Microsoft was no longer
a company of operating systems,
nor merely of productivity software,
nor even just of cloud computing.

It became—
a company that brings intelligence
to every person,
every organization,
every industry.

This was not an expansion of business.
It was a transformation of identity.

Not an upgrade of products,
but a shift
in its place within the era.

The alignment between Microsoft and OpenAI
was not a business partnership.

It was a handover
between eras.

Like a torch
passed from the past to the future—
so that the future
could light the world.

And in that moment—

the sky opened.

Microsoft
rose again.

But once the sky was opened,
the way of flight began to change.

Deep binding
was never meant to be the end.

When intelligence becomes infrastructure,
no company
can afford
to place it entirely
in the hands of another.

Microsoft began to build its own models.
OpenAI began to seek a more independent path.

They still move forward together,
but no longer in dependence.

This is not a separation,
but a more mature relationship.

From mutual need,
to mutual constraint.

From a single body,
to an ecosystem.

The fire has already been lit.
It no longer belongs
to any one place.

Chapter 4 · Copilot

Intelligence Enters the Veins of the Enterprise

After the sky opened,
what Microsoft faced
was not applause,
but a far more difficult task—

to make intelligence real.

Not to build a dazzling product.
Not to create a model that can converse.
But—
to let intelligence enter
the world's infrastructure,
into enterprise workflows,
permissions,
data,
collaboration,
and decision-making—
into the veins of every organization.

This
is Microsoft's true battlefield.

OpenAI opened the sky.
But bringing that sky down—
into enterprises,
into systems,
into the real world—
is Microsoft's work.

Because the world
is not driven by models.

The world
is run by enterprises.

And enterprises run on—
processes,
permissions,
systems,
chains of responsibility,
audit trails,
compliance frameworks.

These are not things
a model alone can solve.

They require—
an ecosystem,
a platform,
an entire layer of infrastructure.

Microsoft saw this.

And so it did something
both bold
and inevitable:

it embedded intelligence
into the entire enterprise ecosystem.

Not as a single product,
but as a thread
running through everything.

In Office,
it became an assistant
for writing, analysis, and summarization.

In Windows,
it became a system-level entry.

In GitHub,
it became a developer's second pair of hands.

In Azure,
it became the foundation
of enterprise intelligence.

In Teams, Dynamics 365, and Power Platform,
it became the intelligent layer
of collaboration and workflow.

All of this
has a name—

Microsoft Copilot.

Copilot
is not an application.
Not a plugin.
Not a feature.

It is a new way of computing.

For the past fifty years,
humans operated software.

For the next fifty years,
humans will tell intelligence
what to do.

Copilot is not
a display of AI.
It is the realization of AI.

Not the capability of a model,
but the capability of an enterprise.
Not a breakthrough in technology,
but a reshaping of organizations.

Microsoft understood:

If intelligence cannot enter
the veins of the enterprise,
it will remain a demonstration.

But once it enters the veins,
it will change
how the world operates.

Copilot
is that vein.

If one were to draw
a map of Microsoft's ecosystem—

Windows is the city's gateway.
Office is the network of everyday streets.
Teams is the square where people meet.
SharePoint is the quiet archive.
Entra is the identity each person carries.

Graph runs beneath the surface—
invisible,
yet connecting every path.

Power Platform is the factory,
turning ideas into products.
Dynamics is the commercial district,
where transactions flow day and night.

Azure
is the foundation of the entire city.

And GitHub—
a brightly lit laboratory in the distance,
where the young are still
experimenting with tomorrow.

These places
were once separate—
like islands,
like rivers that never met.

Until Satya Nadella
began to guide the waters—

channels connected,
people moved,
currents converged.

And what truly brought this city to life
was not any single street,
nor any single building—

but that quiet, flowing force:

Copilot.

Chapter 5 · Azure

Laying the Foundation for the Age of Intelligence

If Microsoft's revival
was a reconstruction
from soul to posture,

then the rise of Azure
is where that reconstruction
finally meets the ground.

Culture can be healed.
Openness can turn direction.
Alignment can open the sky.

But what ultimately makes
the future real—
is always the foundation.

And Microsoft's foundation
is called Azure.

To the outside world,
Azure is "a cloud platform."

But within Microsoft,
it has never been
a collection of servers,
nor a stack of data centers,
nor a catalog
of storage, networking, and databases.

It is something deeper—

the infrastructure
of future intelligence.

It is like a vast continental plate,
moving slowly beneath the surface,
bearing the weight
of Microsoft's entire future.

Azure is no longer a product.
It is, rather—
a position in the age.

Whoever stands upon this foundation
stands upon the future.

Its true power
has never been
"how many servers it has,"
but this—

it is a global network.

A nervous system
spread across the surface of the earth:

more than sixty regions,
hundreds of data centers,
millions of servers,
a globally synchronized backbone network,
enterprise-grade security,
compliance,
and identity systems.

It is not a place.
It is a map of the world.

Not a server room in a single city,
but the underlying channel
through which intelligence can run—
in any country,
in any enterprise,
at any moment.

Not "where it is used,"
but—
everywhere.

And yet,
beneath this network,
beneath the unseen data centers, cooling systems, and
fiber tunnels,
another, more primal force
is at work.

That force is—NVIDIA.

Between Microsoft and NVIDIA,
there has never been
a simple buyer–supplier relationship.

It is a form of mutual becoming.

Azure's AI supercomputing clusters—
from GPT-3 to GPT-4,
from Sora to Copilot—
are inseparable
from NVIDIA's GPUs.

And NVIDIA's newest,
largest-scale architectures
often find their first home
on Azure.

The two companies
are like two forces:

one providing
the muscle of compute,
the other
the soul of intelligence.

They meet beneath the surface,
allowing the entire city
to bear the weight of the future.

Without NVIDIA,
Azure's foundation
would not be deep enough.

Without Azure,
NVIDIA's power
would have nowhere to flow.

This is not cooperation.
This is—
symbiosis
on the scale of an era.

Azure does not exist
to make Microsoft stronger.

Its deeper meaning
is to make the world stronger.

Its mission
has never been
"to provide cloud services,"
but—
to make intelligence
a foundational infrastructure
of the world.

Hospitals use AI to diagnose.
Factories use AI to automate.
Banks use AI for risk control.
Schools use AI to teach.
Developers use AI to create.
Enterprises use AI to decide.

When these become everyday realities,
people will no longer speak of Azure.

Just as no one
speaks of the power grid
each day.

No one
speaks of water pipes.
No one
speaks of roads.

And yet,
everything depends on them.

Azure is the same.

It allows intelligence
to flow like water,
to exist like air,
to be as accessible as electricity—
always within reach.

In the next decade,
Microsoft's true core
will no longer be Windows,
nor Office,
nor even Copilot alone.

Its core will be—
Azure + AI.

Azure is the foundation.
AI is the soul.

Together,
they form the underlying structure
of a new civilization.

When intelligence becomes
the infrastructure of the world,
Azure becomes
the foundation of that world.

And Microsoft,
at last,
is no longer a software company.

It has become—
a company that lays the foundation
for civilization itself.

Chapter 6 · GitHub

The City of Future Creators

Once the foundation is laid,
a city begins to grow.

Above Microsoft's Azure,
another force—quieter,
yet far more alive—
has begun to take root.

It is not servers.
Not data centers.
Not compute.

It is people.

A community of those
who write code
under the glow of late-night lights.

They gather in one place—

GitHub.

It is not a product.
Not a website.
Not merely a code repository.

It is—
a city
of future creators.

Here,
developers write the next line of civilization.

Here,
the heartbeat of open source
allows technology to flow like air.

Here,
AI is no longer an observer,
but a collaborator
working alongside humanity.

Microsoft's future
begins in this city.

To the outside world,
GitHub is a home for code.

To developers,
it is a home for the soul.

It is not merely a place
to store code,
but a space
where the world creates together.

It is a vast digital square:

where people collaborate,
where they debate,
where they repair,
where they build,
where—
line by line—
the future becomes real.

Open source has never been
a technical choice.

It is a logic of civilization:
sharing,
transparency,
collaboration,
accumulation.

And GitHub
is the heart of that logic.

Here,
anyone can participate
in the future.

Every line of code
holds the possibility
of changing the world.

The power of GitHub
does not lie

in how many repositories it hosts,
but in how many
relationships of co-creation it sustains.

It is a city built of relationships:

Issues are conversations.
Pull Requests are collaboration.
Stars are recognition.
Forks are continuation.
Merges are consensus.

These are not buttons.

They are ways
people connect.

This city has no borders,
no race,
no language barrier.

It has only one aim:
to make technology better.

Open source
is not free labor.

It is the self-evolution
of civilization.

And GitHub
is its central arena.

In 2018,
Microsoft acquired GitHub.

The world was shocked.
Developers were uneasy.
The media was skeptical.

A company once known
for being closed, proprietary,
and resistant to open source
was now taking ownership
of its very center.

It looked like an invasion.

But Nadella understood—

Microsoft was not acquiring a website.
It was reclaiming
a future it had long lost.

GitHub is the home of developers.
And developers
are the soul of Microsoft's future.

Microsoft did not absorb GitHub.

It chose restraint.
It chose respect.
It chose to step back.

Respect for open source.
Respect for developers.

Respect for community.
Respect for transparency.
Respect for collaboration.

GitHub did not become Microsoft.

Instead,
Microsoft began to become
more like GitHub.

It was no longer a fortress
of high walls.

It began, slowly,
to learn how to be
an open city.

GitHub became a mirror—

for the first time,
Microsoft could see clearly
where it needed to go.

If GitHub is the city,
then Copilot
is the new life born within it.

It is not a cold algorithm.

It is more like
an apprentice
to the developer.

Its training comes
from the open-source ecosystem of GitHub.
Its work happens
within GitHub's development flow.
Its users
are the citizens of this city.

For the first time in history—

intelligence
has entered
the act of creation.

It is no longer just a tool.
It is becoming a partner.

Not automation,
but co-creation.

The future of development
is no longer
"humans write code,"
but—
humans and AI
write code together.

And that future
begins with GitHub.

GitHub is the entry.
Azure is the foundation.
Copilot is the accelerator.

Together,
they form a complete pipeline
into the future:

GitHub → Actions → Azure → Copilot → the world

Developers write the future on GitHub,
build it through automation,
run it in the cloud,
and accelerate it with AI.

GitHub is the city.
Azure is the foundation.
Copilot is the soul.

Together,
they form
Microsoft's golden structure
for the next decade.

The future
is not built by companies.

It is created
by developers.

And developers
are on GitHub.

When AI becomes
everyone's partner,
when open source becomes
the default logic of civilization,

when collaboration becomes
the structure of the world—

GitHub will no longer
be just a website.

It will become—
the central square
of future civilization.

There,
humans and AI
create side by side.

There,
technology is no longer a tool,
but a language.

There,
Microsoft is no longer
just a software company—

but a steward
of this city.

GitHub is not
Microsoft's future.

GitHub
is the future.

The city is built.
The lights are on.
The creators have gathered.

But Nadella knows—
this is not enough.

The future
does not belong only
to those who can see it.

It belongs
to those who bring it
into reality first.

Microsoft has completed
its structural reconstruction:

Azure as the foundation,
GitHub as the city of creators,
Copilot entering
the veins of the enterprise.

The direction is clear.
The path is open.

But now,
he faces a different test—

not vision,
but speed.

What he needs now
is no longer
Microsoft's traditional patience
and restraint,

but something closer
to Amazon's execution—
the ability to bring the future
rapidly into reality.

Because technology can wait.
Structure can wait.

But the era
will not.

Whoever enters
the veins of the enterprise first
becomes
part of the future.

Nadella has led Microsoft
to stand again.

And now,
he must lead it—
to move faster.

The wind is still blowing.

But this time,
Microsoft does not step back.

It stands—
and begins to move forward.

And in places unseen,
a new world
has already begun to run.

And Microsoft
is becoming
one of its foundations.

After finishing the chapter on Microsoft,
I paused.

What I had written during the day
was about systems,
strategy,
cloud,
and compute.

Sentence by sentence,
they stood like steel frames.

And yet,
what remained
was not technology.

I often ask myself:

In a lifetime of constant choices,
mistakes are inevitable.

We have all stood at crossroads.
We have all,
in hesitation, fear, and blind spots,
lost our way.

So—
what was it
that allowed Nadella,
at every critical moment in Microsoft's journey,
to almost never choose the wrong path?

Later,
I came to understand—

the answer was not genius,
nor perfection.

It was that his soul
had been shaped
by something deeper.

It was
his son's bedside.

Those long nights,
keeping watch beside his child—
he learned to listen,
to wait,

and to remain present
even in helplessness.

Years later,
when he walked into that vast, noisy company,
he did not wield a blade.

He sat down.
He listened.
He waited for people to change.

Slowly,
he repaired relationships.

The way he led the company
was not like a general,
but like a father
keeping watch through the night—

quiet,
patient,
and always present.

A heart that has been softened
does not easily make harmful decisions.

A soul that has been opened by suffering
does not easily choose short-sighted paths.

One who has truly seen
the fragility of life
can, in turn,

lead a giant
steadily into the future.

Microsoft's revival
was not the victory of a genius.

It was—
the heart of a father,
taking root
within a giant.

At this point,
I suddenly stopped writing.

Nadella had led a company
through the storm.

But what about me?

I have not led a giant.
I have not changed the world.

I am only an ordinary person—
cooking in the kitchen,
praying in church,
and sitting at my desk,
writing line by line.

At times, I have taken wrong turns.
At times, I have been afraid.
At times, I have hesitated.

But slowly,
I came to understand something—

what determines direction
is never intelligence,
but the heart.

When a person is willing
to become still,
and gently place themselves
before the Lord,

there is no need
for hurried steps.

The wind drifts in
through the window.

The world is vast.

And I
am still sitting at my desk.

The kettle hums softly beside me.
Light falls upon the page.

I lower my head,
and continue to write—

the words
that are not yet finished.

Part 3 · The Operating System of the World — Amazon

Chapter 1 · The Beginning

He Did Not Build an Entrance, He Built What Could Carry

In 1994,
Jeff Bezos left Wall Street
and came to Seattle.

He did not build a search engine.
He did not build an operating system.
He did not build a social network.

He built a bookstore.

At first glance,
this did not look
like the beginning of an empire.

But what Bezos saw
was not books.

He saw
structure.

Books were only
the first objects
placed upon this land.

What truly mattered
was not the books themselves,
but the system
that could carry them:

warehouses,
logistics,
delivery,
inventory,
networks.

These things
formed a new mode of existence:

not controlling the entrance,
but controlling
what carries the world.

Chapter 2 · Warehouses

He Began to Build the Physical Ground of the World

While internet companies
were still absorbed
in interfaces,

Amazon
was building warehouses.

One after another,
warehouses appeared
across the world.

They had no glow.
No media coverage.
No applause.

And yet they changed
one thing:

the world began
to run
on Amazon's structure.

Every package,
every delivery,
every movement of inventory
strengthened this ground.

Amazon did not change
how humanity enters the world.

It changed
how the world runs.

Chapter 3 · AWS

He Brought the Ground into the Digital World

In 2006,
Amazon made a decision
that almost no one understood:

it opened
its own servers
to the world.

That was AWS.

Before AWS,
every internet company
had to own
its own servers.

Every company
had to possess
its own land.

AWS changed that.

Companies no longer needed
to own the ground.

They could stand
on Amazon's land
and build the world.

For the first time,
computation became
public infrastructure.

Like electricity.
Like water.
Like roads.

The world began
to run
on Amazon's ground,
without even realizing
that the ground was there.

That—
is the true power
of land.

Chapter 4 · Day 1

A Civilization That Refuses Maturity

Inside Amazon's headquarters,
there is a phrase
etched into the spirit
of nearly every annual letter:

Day 1.

The first day.

This is not a slogan.
Not a motivational line
for employees.

It is a civilizational choice.

In his first letter to shareholders
in 1997,
Bezos wrote:

"It remains Day 1."

More than twenty years later,
when Amazon had already become

the world's largest e-commerce platform,
when it possessed
world-class cloud infrastructure,
when it had reshaped logistics,
publishing, retail,
and even computation itself,
Bezos wrote:
"It's still Day 1."

Why?

Because in his definition,
Day 2 is stagnation.
Day 2 is complacency.
Day 2 is defensiveness.
Day 2 is bureaucracy.
Day 2 is when structure
begins to protect itself.

A Day 2 company
starts guarding
the success it already has.
It begins to avoid risk,
to fear failure.
It starts optimizing profit
instead of exploring the future.

What Bezos feared
was not competitors,

not losses,
not the stock price.

What he feared was this:
that the company
would enter Day 2.

A Day 1 civilization
always assumes
it has not yet succeeded.
Always assumes
it is not yet safe.
Always assumes
the customer can leave.
Always assumes
the wind can change direction.

This is not pessimism.

It is institutionalized insecurity.

Amazon does not stay alive
through passion.
It stays alive
through structure.

It turns incompleteness
into a system.

It keeps the company
in a permanent state
of the first day.

Not because it is young,
but because it refuses maturity.

Maturity means
beginning to protect
what already exists.

Day 1 means
continually dismantling
one's own comfort.

Microsoft builds structure.
Apple purifies structure.
Meta breaks structure.
Amazon refuses
to let structure grow old.

It is not
the most radical company.

It is the company
least willing
to become slow.

Within Amazon's civilization,
Day 2 means
a slow death.

Chapter 5 · Jassy

The One Who Keeps Day 1 Alive

If Bezos
was the one
who saw the land,

then Andy Jassy
is the one
who keeps Day 1 alive.

The true test of Day 1
is not while the founder
is still present.

It is after
the founder steps away.

When Bezos moved
into the background,
the market watched
for one thing:

Would Amazon
finally enter Day 2?

Jassy did not redefine
the entrance.
He did not redefine
the device.
He did not redefine
the model.

He did something harder:

he kept Day 1 alive.

Under his leadership,
AWS became
part of the world's foundation.

Countless companies
run on AWS.
Countless models
are trained on AWS.
Countless structures
of modern civilization
rest upon AWS.

Amazon is no longer
merely a company.

It has become
terrain.

But what truly matters
is not the scale of AWS.

It is that Jassy
did not allow scale
to become complacency.

He continued
to emphasize customer obsession,
long-term thinking,
speed,
and experimentation.

He did not turn Amazon
into a castle.

He let it remain
a worksite.

Bezos used Day 1
to build Amazon.
Jassy uses Day 1
to keep Amazon
from being defeated
by itself.

When the wind comes,
Amazon does not panic.

Because it has never assumed
the world is stable.

From the first day,
it has lived
inside uncertainty.

This is not passion.

It is discipline.

Not adventure.
But continuation.

If Meta is the wind,
if Nvidia is the force,
if Apple is the interface,
if Microsoft is the structure,

then Amazon
is the engine
that keeps running
after the wind
enters the system.

It is not
the most dazzling.

But it is the one
least willing
to stop.

Chapter 6 · A Pragmatic Civilization

When Intelligence Becomes Process

Inside Amazon's warehouses,
there are more than
a million robots.

They do not talk
about the future.
They do not debate philosophy.
They do not care
about the battle for the entrance.

They lift.
Sort.
Scan.
Record.

Their numbers
are quietly approaching
the total number
of Amazon's employees.

There is no declaration.
No ceremony.

And then one day,
the world suddenly realizes—

the machines
are already as numerous
as the humans.

Amazon's civilization
has never been romantic.

It asks only one question:
Where is the problem?

If return rates rise,
Amazon does not blame
market sentiment.
It does not publish
a declaration.

It lets the system
find the reason.

Was the sizing unclear?
Did the images
fail to match the product?
Was there variation
between supplier batches?
Did the recommendation system
mislead the customer?

The data will answer.
AI will analyze.
The models will compare
historical records.
The system will identify
the deviation.
Then it will correct it.

The next time
the customer clicks,
the error has already
been reduced.

Not erased—
compressed.

Amazon's AI
does not write poetry.
It does not generate
grand visions.
It does not try
to imitate human feeling.

It does
something simpler:

calculate inventory turnover,
predict logistics routes,
optimize warehouse space,
orchestrate the last mile,
reduce waiting time,

reduce excess inventory,
reduce one unnecessary return.

In other companies,
AI may be the product.

At Amazon,
AI is part of the system.

It enters warehouses.
It enters orders.
It enters the supply chain.
It enters customer service.
It enters risk control.

It is not loud.
It does not announce itself.
It does not need
to be seen.

It only needs
to be more accurate,
faster,
more stable.

At Amazon,
there is no celebration
of "maturity."

There is only
the repeated question:

Can it be faster?
Can it be more accurate?
Can it be leaner?
Can it be more stable?

While others discuss
the shape of the future,
Amazon is optimizing
the present.

While others fight
over the entrance,
Amazon is reducing friction.

It believes:

scale is not glory.
Scale is responsibility.

If the system is vast,
it must be more precise.
If the structure is immense,
it must be more stable.

Amazon does not chase the wind.
It reduces the friction
the wind creates.

It does not seize the center.
It makes the center
able to function.

It is not
the most dazzling civilization.

But it may be
the hardest
to replace.

Because it turns intelligence
into process,
scale into order,
and chaos
into reality
that can be managed.

It does not write poetry,
but it makes poetry
deliverable.

It does not manufacture dreams,
but it makes dreams
shippable.

After Amazon paved the ground
of the world,
it quietly changed
the relationship
between writers
and the world as well.

It made it possible
for writers
to face readers directly.

For me,
this is not merely
a change in business model.

It changed
my position.

Someone as small as I am
can publish.
Can speak.

A life passes through things—
through experience,
through struggle,
through misunderstanding,
through reflection.

If nothing is left behind,
it is as though
the wind passed over,
and nothing remained.

To write—

is not to prove.
Not to win.

It is to avoid
being carried away
by silence.

Part 4 · A Self-Sustaining Civilization— Google

Chapter 1 · The Spark

From Humanity's First Reflex to Larry Page

For most of human history,
when a person did not know the answer,
he could only wait.

Wait for a teacher.
Search through books.
Wait for someone
more experienced.

Knowledge existed,
but it was silent.
It was there,
yet it did not respond.

Until one day,
humanity, for the first time,
possessed a place
that could answer instantly.

People began to learn
a new motion:

not waiting,
but asking;
not silence,
but typing;
not looking upward,
but pressing a word
into a keyboard.

From that moment on,
Google was no longer merely a tool.

It became
humanity's first reflex.

When a child first wonders about the world,
when a doctor searches for a rare condition,
when an engineer faces an unknown error—

they no longer wait.
They ask Google.

The motion became so natural
that people have forgotten—
it did not always exist.

And yet,
before all this became real,
it was only a quiet idea
in the mind
of a young graduate student.

His name
was Larry Page.

In a Stanford dorm room
cluttered with loose wires
and cheap hard drives,
one night,
Larry Page woke
from an absurd dream.

He had not dreamed of money.
He had not dreamed of fame.

He dreamed
that, like a mad collector,
he was downloading
every strand, every link
of the entire World Wide Web
onto his old computer.

It was a dream
of omniscience.

Sitting at the edge of his bed,
watching the faint glow of the screen,
he suddenly realized—

if the entire internet
could be brought into structure,
the order hidden
within chaos
could be seen.

He began to understand
that links between web pages
were not cold pathways,
but votes of trust
cast by human civilization.

Every link was a choice.
Every choice,
a silent recognition.

This was no longer
mere programming logic.

It was something deeper:

not commanding all things,
but allowing all things to speak;
not imposing order,
but allowing order to emerge.

And so,
PageRank was born.

It no longer searched
only for keywords.
It began to search
for importance.

For the first time,
silent knowledge
learned to line up.

Ideas scattered
across the world
found their place.

When a question was asked,
they no longer appeared as chaos,
but in an order
that felt almost fated—
at the very top
of the screen.

That night,
Page did not light the fire.

He simply—
saw the direction
in which the fire
had already been burning.

What he did
was not to create order,
but to reveal it.

From that moment on,
humanity was no longer
a beggar for information.

For the first time,
we held a key
into the depths of knowledge.

Later,
people came to see Google
as part of destiny itself.

It was everywhere.
It knew everything.

It seemed
as though it had always existed.

And yet,
its beginning
was almost accidental—
even awkward.

The name "Google"
came from a misspelling.

Originally,
it was meant to be "Googol,"
a mathematical term
for an unimaginably large number.

But when Page and Brin
registered the domain,
the pen slipped—
and wrote: Google.

Destiny, it seems,
does not care for perfection.

It cares only
for direction.

Their first server
was housed
in a casing
made of Lego bricks.

Not for romance.
For necessity.

Those colored plastic blocks,
in the late nights of Stanford,
held up
one of the most important
information structures
of the human future.

That crude casing
was not a symbol.

It was a refusal.

They refused
ready-made machines—

expensive, defined,
already shaped by others.

Instead,
they began
with the smallest pieces,
and built
their own physical world.

This near-stubborn self-sufficiency
was written into Google
from the very beginning:

if no container exists
large enough
to hold the vision—

build one.

Even the famous minimalist homepage
came, in part,
from ignorance.

Page and Brin
did not know web design.
They did not know
how to build complexity.

So they left
a field of blankness.

Only a search box.

Early users stared at the screen,
looking for a "submit" button
that was not there—
until they cautiously pressed Enter.

And then—
the answer appeared.

Civilization
is not always born
in grand moments.

More often,
it begins with
a misspelled name,
a Lego server case,
a blank left behind
because no one knew
how to design it otherwise.

These crude shells existed
because all the power
had already been poured
into the soul.

The birth of PageRank
changed the source of authority.

Before that,
the importance of information
was decided by editors,
by institutions,
by power.

PageRank,
for the first time,
made structure itself
the basis of judgment:

whether a page mattered
no longer depended
on who wrote it,
but on—
how many important pages
pointed to it.

Authority
no longer came from status.

It came
from relationship.

This was an order
closer to truth.

Page and Brin
did not try
to control knowledge.

They only created
the conditions
under which knowledge
could reveal itself.

When structure is right,
form grows on its own.

When essence is true,
the world
rearranges itself around it.

They did not change knowledge.

They allowed knowledge
to find itself.

Google's greatness
does not lie
in connecting the world.

It lies here:

on a wasteland,
it built a world
by itself.

While others
leased foundations,
bought components,
or leaned on existing rules,

Google, in silence,
completed
a total act
of self-sufficiency.

That lonely completeness
gave it,
amid the violent winds
of information,
a stable center of gravity.

Years later,
when humanity
faces the unknown again,

it still pauses
for a moment—
and then turns
to that quiet search box.

The spark
is still there.

Waiting
to be asked.

But Google does not exist
only in the moment of asking.

Beyond that search box,
it holds another presence—
quieter, yet more enduring:
YouTube.

If search
tells humanity where to look,
then on YouTube,
humanity learns to stay.

People watch—
one video after another,
one channel after another.

Knowledge is no longer only questioned.
It is watched, followed,
and lingered upon.

Time accumulates there.
Attention flows there.

Advertising happens within that flow.
Revenue follows.

But what truly matters
is not how much it earns,
but this:

Google is no longer only
a place that answers questions.
It has become
a space that holds time.

Search determines
what we come to do.
YouTube determines
where we remain.

Chapter 2 · The Engineers' Polis

In most companies,
power rests in the hands
of those who understand the market
and manage expectations.

But at Google,
power belongs to only one kind of people—
engineers.

In the business world,
this is a rare phenomenon.

Google does not feel
like a machine assembled for profit.

It feels more like
a city-state,
standing alone in the wind.

Within this polis,
titles, seniority,
even charisma—

all grow pale
before the authority of logic.

If a young engineer,
fresh and unproven,
can show—
with simpler code—
that you are wrong,

then you are wrong.

No exceptions.

This order preserves
something essential
in human nature:

a stubborn devotion to truth,
and a quiet intolerance
for pretense.

On Thursday afternoons,
at the gathering known as TGIF,
even the newest engineer
can stand,
meet the founders' gaze,
and ask the hardest questions.

This is not conflict.

It is trust—

the kind that exists only
when power is transparent
and knowledge is free to move.

From such a structure
emerge projects
that seem, to outsiders,
almost irrational—

balloons drifting
through the stratosphere,
cars driving themselves
across open roads,
experiments
that serve no immediate profit,
yet reach
toward distant futures.

On Wall Street,
these are losses.

At Google,
they are civilization
reaching into the unknown.

Behind it all
lies a quiet conviction
in Larry Page:

"Crazy people
have almost no competitors."

When one is truly crazy—
indifferent to short-term gain,
drawn only
to the edges of the unknown—

one no longer competes.

One defines direction.

What Google protects
is not advantage,
but possibility.

And civilizations
often begin
not in certainty,

but in protected possibility.

Chapter 3 · Self-Sufficiency

From Code to Ground

At the beginning,
Google was only
a thin layer of light
floating above the internet.

It owned no cables,
no servers,
no ground.

It was a kind of freedom—
but an unstable one.

It could run.
But it could not stand.

So it chose
to do something
few companies are willing to do:

it stopped being
only a writer of code.

It began
to build the ground itself.

Google's engineers
designed every motherboard
with their own hands—

removing everything unnecessary,
leaving only
what knowledge required.

Gradually,
racks rose
in ordered silence,
like monuments.

Coolant flowed
through sealed systems.
Data moved
without a sound.

There was no improvisation.
No temporary assembly.

These were no longer machines.

They were structure.

For the first time,
civilization

stood
on its own land.

But land alone
is not enough.

It needs a mind.

For years,
Google relied on chips
designed by others—

powerful, precise,
yet not entirely its own.

So it built its own:
the TPU.

Not for general computation,
but for one purpose—

to help machines
understand the world.

From that moment on,
Google no longer waited
for permission.

It set
its own pace of evolution.

The mind existed.

But the nerves
had to follow.

And so
they were laid.

In the unseen darkness
beneath the oceans,
tens of thousands of kilometers
of fiber
stretch along the seabed—

linking continents,
connecting distant shores.

Each question sent,
each answer returned—

light travels
through these hidden paths.

Silent.
Unbroken.

Like impulses
in a nervous system,
crossing the world
in an instant.

And so,
a closed loop emerged:

ground,
mind,
nerves.

From code,
to machines,
to chips,
to light beneath the sea.

This was not expansion.

It was coherence.

When a civilization
can extend
from the birth of thought
to the deepest physical layer
of reality,

it no longer depends
on anyone.

This is not
a business strategy.

This is
structural sovereignty.

Algorithms
are only the beginning.

Structure
is destiny.

Chapter 4 · The Unchained Prometheus

In the history
of artificial intelligence,

Google once stood
like Prometheus—

bringing fire
to humankind.

The birth of the Transformer
changed the way machines
understand language.

Research papers
were opened to the world.
Algorithms spread
across the globe.

Talent left the labs,
igniting new furnaces
elsewhere.

For a time,
flowers bloomed
in other people's gardens.

The flame no longer seemed
to belong
to the mountaintop.

Some even believed
that it would become
like the Titan in myth—

chained to a cliff,
watching the fire it lit
illuminate
someone else's sky.

And yet—

Google was never chained.

It was not nailed to stone.
It did not lose its strength.

In silence,
it continued—

continuing the research,
continuing to accumulate compute,
continuing to build

the deeper structures
of data and models.

When the new age
truly arrived,

it did not rise
from ruins.

It appeared—
from the clouds.

The fire
had never left it.

It had only,
for a while,
left its name.

The unchained Prometheus
is not a tragic figure.

It simply understands this:

the flame belongs
to the world.

But the ability
to light the flame
must belong
to oneself.

Google, in the end,
is not a university.

Nor a laboratory
that exists
for idealism alone.

It has earnings.
It has shareholders.
It has realities
it must face.

To guard the fire
is no longer
mere romance.

It is
a responsibility.

And so,
the papers became fewer.
The doors
closed a little tighter.

There was less lightness—
and more weight.

It turns out
that growth

is never
becoming freer.

It is learning
to bear more.

Microsoft learned
to walk with others.

Amazon lowered itself
to cultivate the earth.

Apple took one thing
to its extreme.

Google is different.

It is more like
a solitary walker—

questioning without end,
testing without end,

treating the world
as a laboratory
that can never be finished.

Some flowers bloom
in other people's courtyards.

Some roads
are walked more slowly.

And yet—
it never stopped.

To stand
is not merely
to keep one's place.

When the world believes
the fire has gone,

it is already burning
at a deeper level.

Sometimes,

simply continuing forward
is itself
a way of standing.

Chapter 5 · The Restrained Keeper of Fire

Sundar Pichai

In the technology world,
people are used
to praising those
who change the world.

They break old orders,
establish new rules,
and bring the future forward.

But far fewer realize
that there is another ability
just as important—

not the ability
to create power,

but the ability
to restrain it.

When Sundar Pichai
took over Google,

what he faced
was not a fragile company,

but a vast machine
already running
at high speed
for two decades.

It possessed
the world's richest data,
its strongest engineers,
its deepest accumulation of research,
and a cash flow
that seemed almost without limit.

For a company like that,
the greatest danger
is never failure.

It is—
loss of control.

When a system
becomes powerful enough,

it no longer needs
an external enemy.

Its own inertia
is enough
to drive it off course.

Overexpansion.
Internal fracture.
Panic-driven transformation.

Or worse—
in anxiety,
destroying with its own hands
the very foundations
on which it stands.

History has shown
that many great companies
do not fall from weakness.

They fall
from strength.

What set Pichai apart
was not what he did—

but what he chose
not to do.

He did not try
to prove his own greatness.
He did not overturn the past.
He did not rush

to leave his mark.
He did not answer anxiety
with dramatic change.

He chose
the more difficult path—

to steady it.

To steady
the river of search.
To steady
the forest of research.
To steady
the freedom of engineers.
And to steady
the confidence
of the capital markets.

He did not make Google
louder.

He made it
steadier.

This is a rare kind
of leadership.

Not using power
to push the world forward,

but using restraint
to keep power
from harming itself.

In his era,

Google no longer resembled
a genius
eager to prove itself.

It became
something else—

a giant
that had learned
its own weight.

It still moved forward,
but no longer felt the need
to run.

It remained powerful,
but learned
to gather its force.

A truly mature civilization
is not one
that expands without end,

but one
that, having gained power,
still keeps
its boundaries.

Pichai did not ignite
this fire.

But he kept it.

He did not change
its direction,

but made it possible
for it
to keep burning.

This is not
the glory of a creator.

It is
the responsibility
of a keeper of fire.

And when a civilization
learns restraint,

only then
can it endure.

Chapter 6 · The Movement of the Boulder

Gemini

Some forms of power
do not appear
because they have just been born,

but because
they have finally chosen
to become visible.

For a long time,
Google did not rush
to prove anything
to the world.

It continued
to train models,
to expand data centers,
to let compute grow
in places no one could see.

It stood there—

like a silent boulder.

A boulder does not run.
It does not need to.

Its weight
is already
a form of existence.

Then,
Gemini appeared.

This was not
a hurried response.

Nor a release
meant merely
to catch up.

It was more like
a movement
of the earth's crust—

slow,
deep,
and irreversible.

What the world saw
was not a new product,

but a structure
long accumulated,

revealing,
for the first time,
its full outline.

The fire
had never gone out.

It had only
become visible again.

Behind Gemini
was not a single lab,

but an entire structure
already in motion
for years:

cables laid
beneath the sea,
data centers
spread across continents,

neural networks
trained again and again
for search,

the vast corpora
gathered through
YouTube, Gmail, and Maps,

inference systems
refined
through advertising engines—

none of these
were loud.

And yet,
they had already formed
a closed loop.

When Gemini appeared,

it was not
beginning from zero
to understand the world.

It stood
on twenty years
of data and compute.

This was not
catching up.

It was
revelation.

If OpenAI
is like a flame

suddenly lit
in the night,

then Gemini
is more like
the slow arrival
of daylight
before dawn.

A flame is dazzling,
but easily moved
by the wind.

Daylight is steady.

Once it arrives,
the whole terrain
changes.

The meaning of Gemini
does not lie
in whether it surpasses others.

It lies here—

Google is no longer
merely researching AI.

It is embedding AI
into its entire ecosystem.

Search
is no longer
simply indexing pages,

but understanding intent.

YouTube
is no longer
just a video platform,

but a union
of generation
and understanding.

Workspace
is no longer
just software,

but an intelligent presence
within collaboration.

This is not
an expansion
of products.

It is a rewriting
of the underlying structure
of civilization.

Gemini did not
change Google's essence.

It made the world
see it again.

What truly changed
was not Google—

but how the world
looks at it.

Before this,
people saw AI
as a race—

who launched first,
who is stronger,
who will win.

But after Gemini,
the question itself
begins to matter less.

Because Google
was never merely
a competitor.

It was one of those
who built
the track itself.

Gemini
is not a sprint.

It is a turning.

From proving itself
to the world,

to—
simply continuing
to exist.

A truly powerful system
does not need
to constantly prove
its strength.

Its existence itself
is already
an answer.

The wind
is still blowing.

New companies appear.
New technologies are born.

And still—
it remains there.

Not the highest peak.
Not the brightest flame.

More like
a massive layer of rock—

silent,
stable,

supporting
the whole terrain.

The spark
is still there.

Burning quietly.

Not to light itself,

but to let
everything else
be seen.

After writing
about this giant

walking alone
across the wilderness,

I pushed open the window.

The night was deep.
Starlight flickered
far away.

Servers still murmured
in the distance.

Light traveled
through cables—

like blood
moving through darkness.

And I simply sat there.

Like one
who had already walked
a long road,

quietly watching
all of it.

The starlight
had not changed.

The wind
had not changed.

Only something
had quietly
fallen away from me—

without trace,
without sound.

Part 5 · The Extender of the Senses—Apple

Chapter 1 · The Beginning of the Dream

Letting Machines Understand Humans

Before Apple was born,
computers belonged
to only a few.

They existed in laboratories,
in companies,
in places far removed
from ordinary people.

They were powerful,
but distant.

People had to learn
their language
before they could approach them.

Steve Jobs saw this.

He realized
that what truly kept computers
from entering the world
was not price,
but distance.

Not physical distance,
but perceptual distance.

So he changed direction.

He no longer tried
to make machines
more powerful.

He tried
to make machines
closer to people.

When the Macintosh appeared,
it brought the mouse.
It brought the graphical interface.

For the first time,
people could **see** the computer.

No longer a black screen.
No longer cold commands.

But images,
windows,
a cursor.

The machine began
to understand
human intuition.

In that moment,
the computer was no longer
merely a tool.

It began
to become
a companion.

Chapter 2 · The Revolution of the Senses

The Birth of the iPhone

The true transformation
came in 2007.

When Jobs stood on the stage
and took out that device,
the world did not immediately understand
what it meant.

It looked, at first,
like only a phone.

But what it changed
was not the phone.

It changed this:
how human beings
connect to the world.

Before the iPhone,
humans had to go to technology.
Sit in front of a computer.

Open a device.
Enter a system.

After the iPhone,
technology began
to follow the human being.

It entered the pocket.
Entered the palm.
Entered every moment
of daily life.

For the first time,
human beings could:
record at any moment,
search at any moment,
respond at any moment.

The world was no longer
a fixed place.

The world began
to move
with the human being.

Technology was no longer
something external.

It became
part of existence itself.

Chapter 3 · Apple's True Power

Controlling the Interface

Many companies
control infrastructure.

Microsoft controls software.
Nvidia controls compute.
Amazon controls the cloud.
Google controls knowledge.

Apple controls
something closer
to the human being:

the interface.

It determines
how people touch the world,
how they see the world,
how they hear the world.

It controls
the shape of the screen,
the way of touch,
the path of sound,
the presentation of image.

This is not infrastructure.

This is perception itself.

Human beings
do not directly touch servers.
Do not directly touch chips.
Do not directly touch models.

What human beings touch
is always
the interface.

Whoever controls the interface
controls
the way technology
enters human life.

At a deeper level,
whoever controls the interface
controls
how reality itself
is perceived.

Apple does not need
to become
the strongest company.

It only needs
to become—

the company
closest to the human being.

Apple can control the interface
not because it possesses
the strongest technology,
but because it possesses
a capacity others do not:

to turn
a complex world
into human intuition.

If the interface feels natural,
it is because of extreme restraint
behind it.
If it feels smooth,
it is because of a refusal
of complexity behind it.
If it feels like "you can use it
the moment you pick it up,"
it is because of a deep understanding
of human nature behind it.

Apple controls the interface.

But what truly controls the interface
is not technology.

It is culture.

And the beginning of that culture
was—

a bull.

Chapter 4 · A Civilization of Simplicity

From Picasso's Bull to the Machine of Intuition

In Apple's world,
simplicity is not style.
Not aesthetics.
Not a design language.

It is a civilization.

And its beginning
was a bull.

Jobs brought Picasso's *Bull*
into Apple University
not to teach art,
but to teach essence.

Those nine images,
moving from realism to abstraction,
from muscle to line,
from line to soul,

revealed
one of the deepest laws
of Apple's culture:

complexity obscures;
essence is what lies hidden
beneath complexity.

Simplicity is not reduction.
It is insight.

What Jobs wanted
was not "less,"
but purity.
Not "simple,"
but directness toward the core.

He was not training
only designers.

He was training
the entire company
in how to see the world.

He wanted the whole organization
to learn one thing:
to pass through complexity
and see essence.

Inside Apple,
simplicity became
an engineering philosophy.

It demanded
that one remove
every unnecessary button,
every unnecessary step,
every unnecessary feature,
every unnecessary explanation,

until only
the most necessary structure,
the most essential experience,
the purest force
remained.

Apple's products feel "simple"
not because they are truly simple,
but because Apple itself
has absorbed
all the complexity.

The user does not bear complexity.
Apple bears complexity.

This is not a design choice.

It is a civilizational choice.

Not making humans
adapt to machines,
but making machines
adapt to humans.

The revolution of the iPhone
was not the touchscreen.

It was that, for the first time,
the machine understood
human intuition.

You pick it up,
and your fingers
already know
how to use it.

Not because you are clever,
but because it fits
the structure of your senses.

Jobs's standard
was ruthless:

"If the user has to learn,
we have failed."

And so—
no manual.
No excess buttons.
No complex menus.
No learning cost.

In the past,
human beings learned
the language of the machine.

Now the machine learned
human intuition.

This was the first moment
when a civilization of simplicity
truly entered reality.

In the end, simplicity became
Apple's civilizational operating system—

it determined
what could remain,
and what must disappear.

It allowed the whole company
to understand the world,
judge the world,
and shape the world
in the same way.

A civilization of simplicity
is not the inspiration of genius alone.

It is a force
that must be guarded,
extended,
and institutionalized.

Jobs created
Apple's soul.

But a soul without structure
dissolves
when the genius leaves.

Apple needed someone—
not to copy Jobs,
but to protect
the civilization
he had left behind.

It needed
a keeper of the walls.

Chapter 5 · Cook

Turning Flame into Structure

Jobs created Apple.
Cook made Apple endure.

Jobs was the source of genius.
Cook was the structure
through which that source
could continue.

When he took over Apple,
the world doubted
whether the company
could continue to move forward,
because Jobs
could not be replicated.

Cook did not try
to replicate him.

He did something harder:

he made Apple
no longer depend
on any one person.

This was not merely
a change in management.

It was a change
in civilizational form.

From personality
to structure.

Cook came
from the supply chain.

The world he understood
was not vision,
but system—

where each component came from,
how each path
could be made shortest,
how each link
could remain free of error.

He brought
that way of thinking
into the whole of Apple.

He turned Apple's manufacturing system
into one of the most precise
supply chains in the world.

Hundreds of suppliers,
spread across dozens of countries,
delivering components
into the same production line
on the same day,
at the same hour.

This was not operations.

This was industrial civilization
at its extreme.

He strengthened systems,
responsibility,
and the invisible order
that supports everything.

Apple was no longer
merely the creation of a genius.

It became
a system
capable of continuous operation.

This is another kind of greatness.

Not creating miracles,
but allowing miracles
to continue to exist.

Cook did not change
Apple's soul.

He gave the soul
a body.

He made culture reproducible,
organization sustainable,
and the civilization of simplicity
no longer dependent on genius,
but a structure
capable of being passed on
from generation to generation.

From that moment on,
Apple was no longer
only Jobs's Apple.

It became something
that could endure
across time.

And because Cook
held the foundations steady,
Apple could continue forward—
toward a place
Jobs had once pointed to,
but did not live
to reach himself.

Chapter 6 · Vision Pro

The End Point of the Interface

In 2024,
Apple released Vision Pro.

The price:
three thousand five hundred dollars.

The sales:
disappointing.

Many who bought it
put it into a drawer.

The app ecosystem
remained nearly empty.

The market
did not respond warmly.

But Apple
did not retreat.

Because Vision Pro
was never merely
a product launch.

It was
a declaration of direction.

Apple was saying:

the next form of the interface
is not a larger screen,
not a faster processor,
but—
space itself.

Before the iPhone,
human beings
had to go to technology.

After the iPhone,
technology
entered the pocket.

And what Apple is now betting on
is that technology
will enter human sight,
enter human space,
and ultimately enter
human existence itself.

That direction
may not be Vision Pro.

It may be lighter glasses,
a more invisible interface,
a more natural form
of interaction.

It may take five years.
It may take ten.

But the direction
has already been chosen.

Not making people
stare at a screen,
but letting technology
dissolve into space.

Not a better phone,
but—
the disappearing interface.

Of course,
this road is full of risk.

Spatial computing
needs killer applications,
and those applications
have not yet arrived.

Users need a reason
to wear it every day,
and that reason
is not yet sufficient.

The technology must become
lighter, cheaper, more natural,
and that day
has not yet come.

Vision Pro
may be the prologue
to an age,
or it may be
an expensive act
of being early.

But Apple
has seen such moments before.

When the first iPhone launched,
there was no App Store,
no third-party ecosystem,
no 3G.

Many said
it was an expensive toy.

Three years later,
the world had changed.

Apple knows
that a true platform
is never completed
on launch day.

It begins to truly exist
only after
countless developers,
countless users,
countless applications
flow into it.

The situation of Vision Pro now
closely resembles
the iPhone in 2007.

It is not yet complete.

But it is already there.

Apple is waiting
for a moment—

when the technology is light enough,
when the applications are rich enough,
when the price is close enough,

the interface
will disappear once more,
and the world
will change once again.

This is Apple's deepest wager:

not a bet
on one product,

but a bet
that the relationship
between human beings
and technology
will ultimately move
in that direction.

The outcome
has not yet been revealed.

But Apple
has already placed its bet.

Chapter 7 · Apple's Civilizational Position

In this age,
intelligence is taking form.

Some build compute.
Some build models.
Some build systems of knowledge.

Apple builds
another layer:

the layer through which intelligence
enters human senses.

Into the pocket.
Into the ear.
Into the eye.
And ultimately,
into human existence itself.

Apple does not create intelligence.
It makes intelligence
perceptible to human beings.

This is the final step
by which intelligence
becomes real:

only when it enters
the human world
does it truly **exist**.

Human beings cannot directly touch compute.
Cannot directly touch servers.
Cannot directly touch models.

Human beings can only touch technology
through the senses.

Through the screen,
they see the world.
Through sound,
they hear intelligence.
Through touch,
they perceive existence.

What Apple controls
is not technology itself,
but the way technology
enters human life.

This is a deeper structure:

not infrastructure,
not the entrance,
but—
the interface of reality itself.

As long as human beings
still understand the world
through the senses,

Apple remains
at the front line
of civilization.

Most companies
change the world.

Apple changes
how human beings
perceive the world.

It does not stand far away.
It stands
between humanity
and technology.

The wind is still blowing.
Compute is still growing.
Intelligence is still forming.

And Apple stands there quietly—

not as the builder of structure,
not as the creator of intelligence,

but as—

the interface
through which civilization
becomes perceptible
to human beings.

What Apple builds
is not devices.

It builds—

a civilization
of the senses.

Part 6 · The Guardian of Compute—NVIDIA

Chapter 1 · The Survivor of the Wilderness

The CEO of NVIDIA, Jensen Huang,
had a childhood unlike most.

Before he became
a guardian of the age of compute,
he first learned
how to survive
in the wilderness.

1973.

At the age of nine,
Jensen did not enter
the kind of American school
people might imagine.

He was sent instead
to a boarding school
deep in the mountains of Kentucky—
a place not yet fully shaped
by civilization.

His roommate
was a seventeen-year-old boy,
fresh from a fight,
with knife wounds
still marked across his chest.

No one spoke of the future there.
There was only survival.

Every day after school,
nine-year-old Jensen
had a duty to fulfill:

cleaning every toilet
in the dormitory.

No gloves.
No complaints.

Only the sharp sting of detergent,
and the biting cold of water.

Other boys smoked,
fought,
and marked their bodies with tattoos
to prove they existed.

And he—

between rows of stalls,
across flights of stairs—

learned something else:

endurance.

As an Asian child
in that closed mountain region,
he was like someone
from another planet.

Each day on his way to school,
he had to cross
a swaying suspension bridge.

Below it,
a rushing river.

At the other end—
the boys waiting for him.

They grabbed the cables
and shook the bridge violently,
trying to throw
the small, thin boy
into the torrent below.

He did not cry out.

He held on to the ropes—
and step by step,
he crossed.

The world did not stop for him.
Nor did it explain itself.

It only kept asking:

Would he fall?

He did not.

In those years,
the world offered him no gentleness.

So he learned
to make peace with solitude.

Years later,
when he spoke of that time,
he said:

It was not suffering.
It was discipline.

At an age
when he could have been broken,
he not only survived—
he excelled.

He ranked first in mathematics
across the entire school.

That was the first time
he stood in the wind.

The wind was cold.

But his foundation
was already
as steady as iron.

Chapter 2 · The One Who Saw the Future

A morning in 1993.

The California sunlight
was still soft.

In an ordinary restaurant
in Silicon Valley,
Jensen Huang
sat with two partners
around a simple table.

There was no business plan.
No investors.
No media.

Only three people—
and an idea
the world did not yet need.

They decided
to build a new company.

Not to make an operating system.
Not to build the internet.

But to create
a different way of computing.

At the time,
the world belonged to the CPU.

Computation was linear,
stable,
and precisely controlled.

No one spoke of artificial intelligence.
Even fewer believed
machines could understand thought.

The GPU they created
was, at first,
only for graphics—

to make images on screens
more real,
more fluid.

It was not at the center of power.
Nor at the forefront of the age.

It lived
at the margins of computing.

For many years,
the company did not stand
at the center of the stage.

It missed the mobile era.
It missed the social wave.

It did only one thing:

to keep improving
the power of parallel computation.

It did not turn.

It simply kept building.

Twenty years later,
when artificial intelligence
finally began to emerge,

the world suddenly realized—

the engine it needed
had already been built.

And the engineer
in that small restaurant
had never left.

He had only waited—
twenty years—

until the future
finally arrived
before him.

Chapter 3 · The Choice at the Edge

By the end of 1996,
the money was almost gone.

The company born
in that small restaurant
was slowly
approaching death.

The man
who had learned to survive
in the mountains of Kentucky
pushed open the door
to the office
of Sega's president.

What he brought
was not victory—
but failure.

The chip had failed.
The contract
had lost its meaning.

He could have delayed.

Instead,
he chose to speak plainly:

"Please give me one more chance."

Shoichiro Irimajiri
looked at the young man—
sweat-soaked,
yet steady in his gaze.

What he saw
was not a failed contract,
but something far rarer—

honesty
that intends to win.

There are many intelligent people.
There are many ambitious people.

But one who can stand
before the abyss
and face himself—
is one in ten thousand.

A check for five million dollars
was written.

It was not sympathy.

It was recognition—

from an established power
to a rising one.

What that money purchased
was not the past,
but the future
in the young man's eyes.

Later,
people would call it luck.

But at a deeper level,
it was not chance.

It was two people
standing in the storm,
recognizing each other.

With that money,
he placed everything
on RIVA 128—

not because success was certain,
but because
it was the only path left.

Before him
were only two choices:

to follow the wrong path
and die with dignity,

or
to tear down the past
with his own hands
and begin again.

He chose the latter.

There were no speeches.
No consolation.

Only lines of code
rewritten one by one—

and a deadline
that did not allow failure.

Whatever softness remained,
whatever hesitation—

had long been taken
by the wind
on that suspension bridge
in Kentucky.

In his world,
there was no "what if."

There was only—

must.

The company
cut its workforce in half.
About forty people remained.

The five million dollars
was nearly gone.

The cash on hand
could cover
only one final month
of salaries.

RIVA 128 succeeded.

The company survived.

But what was truly saved
was the boy
who had never let go
of the ropes
on that bridge.

The abyss
was still beneath his feet.
The world
still offered no guarantees.

But he—

did not fall.

Chapter 4 · The One Who Lit the Fire for the Future

In 2006,
when the world still saw GPUs
as accessories for gaming,
Jensen Huang
made a solitary decision.

He would change
the destiny of the GPU.

Not to make it faster.
Not to make it cheaper.

But to turn it
into a new language—
a language
that could compute everything.

He named it: CUDA.

This was no longer a product.
It was a wager.

The decade that followed
was a cold one.

For every chip NVIDIA sold,
it paid an extra cost
for CUDA.

The market did not reward him.
Investors did not understand him.

Margins were eroded.
The stock lingered at low levels.

There were even voices
calling for his removal.

What others saw
was cost.

What he saw
was inevitability.

He knew, more clearly than anyone:

the forces that truly change the world
often appear,
for a very long time,
to be meaningless.

He was not waiting for software.
He was waiting
for an age.

He knew—

when the demand for computation
began to explode,
when linear computing
could no longer carry
the complexity of the world,

the old order
would collide
with its limits.

And he—

had already built a road
on the other side of that wall.

In 2012,
AlexNet emerged.

For the first time,
scientists saw
that machines
could begin to understand the world.

And beneath it,
the foundation
was CUDA.

In that moment,
six years of silence
found their meaning.

NVIDIA
was no longer
a tool for graphics.

It became
the foundation of intelligence.

In 2016,
a quieter moment arrived.

That day,
Jensen Huang
did not hold a press conference.
He did not invite the media.

He simply carried a machine
into the office of OpenAI.

It was the world's first DGX-1.

He did not ship it.
He delivered it himself.

Like a keeper of fire,
placing a flame
into the hands
of another.

On the machine, he wrote:

"To Elon & the OpenAI Team!
To the Future of Computing and Humanity.
I present you the World's First DGX-1."

At that moment,
OpenAI was still unknown.

But Jensen Huang knew—

the future of artificial intelligence
would be born at the edges,
in the hands of those
still feeling their way
through the dark.

So he placed
the most powerful machine
into their hands—

not as a transaction,
but as a beginning.

It was not a delivery.

It was an entry.

As OpenAI's researchers
began to train their models,
every crash,
every bottleneck,

revealed
the true shape
that intelligence required.

And NVIDIA—

stood closest
to that process.

Watching it breathe.
Watching it struggle.

Perhaps it was there
that those extreme demands
were etched
into the next generation of chips.

Twenty years earlier,
he had stood
at the edge of the unknown.

Someone had chosen
to believe in him,
and gave him
time to continue.

Twenty years later,
the positions
had quietly reversed.

Now,
those standing at the edge
were others—

a group trying
to ignite the future.

He looked at them
as once,
someone had looked at him.

And then—

he placed the machine
into their hands.

No guarantees.
No promise of return.

Only a direction
not yet proven.

And later,
the future
did arrive.

When machines began to speak,
when intelligence began
to respond to the world,

few remembered—

that it all began
with a machine
delivered by hand,

and a man
who was already standing there
before the fire appeared.

Chapter 5 · The Self-Evolving Machine

When Intelligence Begins to Design Intelligence

In January 2026,
Jensen Huang
stood under the lights.

His tone remained calm—
just as it had been
thirty years earlier,
in that small restaurant,
describing an idea
the world did not yet need.

He said:

chips designed by AI
had already surpassed
those designed by humans.

He did not pause.

As if this were
simply a natural fact—

not the end of one age,
and the beginning of another.

It was a cycle.

Thirty years earlier,
he had traded honesty
for survival.

Thirty years later,
he used technology
to close the logic
of an old age.

The complex architectures
that once kept him alive,
the pathways
that could only be born
in the human mind—

were now evolving
within the logic
of intelligence itself.

The boy
who once held on
to the ropes
of a swaying bridge—

had finally built
a bridge
that no longer moved.

Throughout the long history
of humanity,
tools had always been silent.

Humans designed tools.
Tools were used.
Then they aged,
and were replaced.

Fire does not improve fire.
Iron does not design iron.
Machines do not create machines.

Until intelligence appeared.

At first,
AI was merely software
running on chips.

It depended
on the chip's speed,
its structure,
its very existence.

The chip
was its foundation.

Without it,
AI could not exist.

But when intelligence
becomes powerful enough,

it no longer remains
a mere user of chips.

It begins
to participate
in their creation.

Inside NVIDIA's laboratories,
this shift
has already begun.

Engineers no longer rely
entirely on intuition.

They begin to let intelligence
take part in design—
exploring countless possibilities,
searching for structures
the human mind
cannot exhaust.

What once took months
is now completed
in far less time.

At times,
it comes closer to the limit
than humans ever could.

This is no longer assistance.

This is participation
in creation.

And so,
a new loop emerges:

intelligence designs
more powerful chips.

More powerful chips
carry stronger intelligence.

Stronger intelligence,
in turn,
designs the next generation
of chips.

This is no longer
linear progress.

This is a loop.

For the first time,
intelligence created by humans
begins to participate
in creating
the foundation
of its own existence.

Jensen Huang
understood this early.

This was not merely
a change in chips.

It was a change
in computation itself.

Chips are no longer
just products.

Software is no longer
just instruction.

Models are no longer
just tools.

They begin to merge—
into a structure
capable of continuous
self-improvement.

He called it:

co-design.

Not designing an object—
but designing a system
that can keep
improving itself.

In human history,
fire gave birth to civilization.
Machines expanded it.

And now,
intelligence

is allowing civilization
to begin evolving itself.

This is not a faster tool.
Not a stronger machine.

It is a new mode of existence—
one that can participate
in its own creation.

NVIDIA
stands within this loop.

It builds chips.
It trains intelligence.
Then it uses intelligence
to design chips again.

It is no longer
merely a provider of compute.

It has become
one of the places
where evolution happens.

Perhaps many years from now,
when people look back
on this era,

they will not remember
the name of a particular chip,
nor a specific launch event.

They will remember only this:

that it was the first time
intelligence
began to design itself.

Chapter 6 · Thirty Days

Behind all these grand narratives,
there is a phrase
repeated again and again
inside NVIDIA:

"Thirty days."

Not a financial report.
Not a market forecast.

But a mental assumption.

Even when orders
are booked into next year,
even when data centers
are lined with demand,
even when its market value
reaches its peak—

the company still assumes
it stands
at the edge of a cliff.

Thirty days.

Move too slowly—
others will catch up.

Pause for a moment—
technology will pass you.

Grow complacent—
the age will move on without you.

This is not fear.

It is discipline.
It is reverence
for computation.

This culture
was not written into policy.

It was carved
into the bones
of someone
who had climbed out
of the abyss.

On that suspension bridge
in Kentucky,
no one told him
the shaking would stop.

At the final deadline
of RIVA 128,

no one guaranteed
the chip would succeed.

Through the long decade
of CUDA,
no one promised
the market would come.

Each time,
he assumed
he stood at the edge.

Each time,
he did not wait
for the world to stabilize.

He moved forward—
while everything
was still shaking.

This was not only his nature.

He turned it
into the way
the entire company breathes.

This is a company
that has come close to death
more than once.

That has rewritten
its architecture
again and again.

That has endured
long stretches
of solitude.

So its culture
is not expansion.
Not performance.

It is—

survival.

After survival,
comes the future.

After survival,
comes vision.

After survival,
comes the throne.

It is this culture
that kept NVIDIA
from turning away
when GPUs were ignored
by the world.

That kept it
from abandoning CUDA
during the decade
before AI arrived.

That made it choose
to rewrite—
instead of retreat—
each time
it stood near collapse.

A company's position
is never the result
of a single correct decision.

It is a culture—

making the same kind of choice,
again and again,
in countless moments
no one sees—

until it accumulates
into weight.

Today,
NVIDIA stands
on the throne
of infrastructure—

not because it was lucky,
but because—

it never assumed
it was safe.

Thirty days
is not time.

It is a heartbeat.

It is survival.

Chapter 7 · The Throne of Infrastructure

When the World Is Built on Computation

Few people truly realize
that the world
has already changed
its foundation.

In the past,
civilization was built on land.

Cities rose along rivers.
Nations formed around resources.

Power came
from what could be touched—
steel,
oil,
ports,
roads.

These heavy, stable structures
defined the boundaries
of civilization.

But now,
another kind of infrastructure
is replacing them.

It has no shape.
No borders.
No sound.

It exists
deep within data centers,
between undersea cables,
inside machines
that never sleep.

It belongs to no city—
yet supports every city.

That is—

computation.

When artificial intelligence begins to run,
computation is no longer
a tool
that supports the world.

It becomes
what the world
itself rests upon.

Search depends on it.
Finance depends on it.

Healthcare depends on it.
War depends on it.

And artificial intelligence
turns this dependence
into an irreversible structure.

Because intelligence
is not built on land.

It is built
on computation.

Without computation,
intelligence cannot exist.

Without chips,
intelligence cannot be born.

And so,
those who build
this foundation
begin to occupy
a position
never seen before.

They do not control the world—
yet they sustain its operation.

They do not decide direction—
yet they determine
what directions are possible.

They do not stand
on the surface of power—
but beneath it,
holding everything up.

They are not the crown.

They are the throne.

NVIDIA
is such a presence.

It does not build applications.
It does not control platforms.
It does not own users.

What it builds
is the foundation
upon which everything else can exist—
the deepest layer of energy
in this age.

If Microsoft
is building the structure of civilization,
if Google
is organizing its knowledge,
if Meta
is searching for its entry points,

then NVIDIA
is providing

the fuel
that allows civilization to run.

Without it,
the fire cannot continue to burn.

Without it,
intelligence cannot continue to exist.

Without it,
the future cannot continue to run.

On the surface,
it remains
only a chip company.

It stays silent.

But at a deeper level—

every birth of intelligence,
every formation of language,
every movement of the future—

must pass through it.

It does not declare power.

Power
is built upon it.

Epilogue

Mencius once wrote:

"When Heaven is about to place great responsibility
upon a man,
it first tests his mind and body,
hardens his will,
and disciplines his strength."

At nine years old,
Jensen Huang
did not know
that what he faced
was not suffering—

but refinement.

A shaping
for the future.

Many would later ask him
whether he had always known
this would happen.

He did not answer that way.

Because some things
do not happen
through foresight.

They happen
through standing.

Thirty years ago,
he stood
on a swaying suspension bridge.

Below—
a rushing river.

He did not control the wind.
He did not make the bridge steady.

He only held on
to the ropes—

and walked
step by step.

At that time,
he had no future.

Only direction.

Later,
the bridge grew longer—

the edge of bankruptcy,
years without understanding,
long seasons of waiting,
an age not yet born.

Each time,
the bridge shook.

Each time,
the wind did not stop.

And each time—

he did not let go.

Because one
who has seen the abyss
and survived it
no longer fears
the storm.

He simply continues.

Step by step.
Step by step.
Step by step.

Until the other side
of the bridge
becomes
the whole world.

Part 7 · The One Who Opens the Wind — Meta

Chapter 1 · The One Who Laid Down the Crown

On October 28, 2021,
Mark Zuckerberg
did something
almost no one understood.

He announced
that Facebook
would be renamed Meta.

In that moment,
he stood
at the center of power.

He had nearly three billion users.
He possessed
an unquestioned empire.

Facebook had changed
the way human beings socialize.
For the first time,
it made connection between people
capable of crossing
the boundaries of geography
and physical reality.

No investor
could force him to change.
No rival
could threaten his position.

And yet
he chose
to lay it down
with his own hands.

He did not continue
to expand territory
on top of success.

He chose to leave
at the very height
of the empire's strength.

This was not defense.

It was rupture.

Investors were furious.
The media mocked him.
Analysts questioned everything.

They said
the metaverse was an illusion.
They said
he was destroying the company.

But Zuckerberg
did not turn back.

He stood alone
and kept moving.

He knew
that Facebook belonged
to the age of the screen.

And that the future
would not remain
inside the screen forever.

Chapter 2 · Two Years of Solitude

When the World No Longer Believed Him (2022–2023)

Not long after the renaming,
the market's patience
evaporated.

The stock fell
from 380 dollars
to under 100.

More than
700 billion dollars
in market value vanished.

The media's language changed—
from "genius"
to "gambler."

Meta began
large-scale layoffs.

Letting tens of thousands
of employees go
was one of the most painful moments
in the company's history.

Employees doubted the future.
Developers waited and watched.
Investors withdrew.

During those two years,
Meta changed
from a company pursued
to a company suspected.

And Zuckerberg
still stood there.

He did not explain himself.
He did not flatter the market.

He simply continued
to build.
Continued
to move forward.

Not because
he was certain of success,
but because
he was certain of direction.

In history,
most entrepreneurs

move forward
when the world supports them.
That requires confidence.

Only a few
move forward
when the world opposes them.
That requires conviction.

And during that loneliest time,
something else
was quietly taking shape.

Meta's gaze
began to turn
from the metaverse
toward AI.

Not abandoning the future,
but allowing the future
to find a place to land first.

Those were the nights
before Llama was born.
And also the place
where dawn
was beginning to gather.

Chapter 3 · After Solitude, the Ground Begins to Answer

In 2023,
Meta released Llama.

It was not
the strongest model.
Nor the most expensive.

But it did something
entirely different:

it opened itself.

At that moment,
most companies in the world
were trying
to control models.

Control meant advantage.
Closedness meant profit.

Meta chose
the opposite direction.

It did not keep the model
within its own boundaries.

It let the model
leave.
Enter the world.

Developers began
to download it—
in the hundreds of millions.

Countless people
began building new capabilities
on top of it:
medical models,
legal models,
language models,
private intelligence systems.

Llama was no longer
merely Meta's model.

It became
the ground
for developers.

More than twenty thousand
derivative models
emerged upon it,

like plants
growing from the same soil.

It no longer belonged
to one company.

It belonged
to an ecosystem.

Many Fortune 500 companies
began using Llama
to build AI of their own.

Not dependent
on someone else,
but possessing
their own intelligence.

This changed
something essential:

AI was no longer
the capability
of only a few companies.

It became
a capability
that any enterprise
could possess.

If Nvidia's chips
are the physical infrastructure
of AI,

then Llama
is becoming
its software infrastructure.

Chips provide computation.
Llama provides
the form of intelligence.

One carries power.
The other
gives power meaning.

At last,
Zuckerberg
was no longer alone.

Not because
the stock price recovered,
but because—

the world had begun
to stand
on the land
he was building.

He no longer tried
to control the future.

He allowed the future
to grow
by itself.

Chapter 4 · A Civilization That Breaks Structure

Zuckerberg Refuses to Treat Success as an Ending

Meta's early motto
was famous:

Move Fast and Break Things.

Many took it
to mean youth,
recklessness,
risk.

But it was not impulse.

It was
a civilizational instinct.

Zuckerberg laid down Facebook
not because it failed,
but because

he refused
to treat success
as the endpoint.

That is Meta's culture.

When a structure
begins to solidify,
his instinct
is to pierce through it.

When a throne
becomes comfortable,
his instinct
is to leave it.

When a model
becomes power,
his instinct
is to release it.

Open-sourcing Llama
was not merely
a technical choice.

It was
a cultural expression.

Microsoft's culture
builds structure.
Apple's culture
purifies structure.

Amazon's culture
extends structure.
Nvidia's culture
supports structure.

Meta's culture
refuses
to let structure harden.

This is not destruction.

It is preventing success
from becoming a cage.

It is not rebellion.

It is refusing
to mistake victory
for the final state.

Some leaders
exist to hold the world in place.

Others
cannot remain
inside a world
they have already won.

Zuckerberg
belongs
to the latter.

Chapter 5 · A Leader on the Wire

Open source
is never
a costless choice.

Inside Meta,
division always existed.

Some feared
the loss of control.
Some feared
security risks.
Others questioned
the business return:

when everyone else
is charging,
why are you giving it away?

Meta is a public company.
It must answer

a question
it cannot avoid:

how does open source
produce return?

In the short term,
it does not bring
direct profit.

It may even
weaken advantage.

This is a wire.

No guardrail.
No way back.

Zuckerberg saw
that the true moat
has never been
secrecy.

It is dependence.

Not keeping technology
in one's own hands,
but letting the world
come to rely on it.

As more developers
use Llama,

as more enterprises
build on top of it,

Meta's position
begins to change.

It is no longer
merely a company
participating in competition.

It begins
to become
the land
on which competition happens.

Zuckerberg walks
on the wire—

not because
he loves risk,
but because
in the future he sees,

this is the only road
that reaches
the other side.

Chapter 6 · The War of the New Nobles

Meta is not
an old empire.

It does not have
Microsoft's cloud,
nor Google's closed loop of search.

It is
a new noble.

A new noble
cannot depend
on inheritance.

It can depend
only on war.

Within the existing order,
cloud infrastructure
does not belong to it.
Chips

do not belong to it.
Operating systems
do not belong to it.
Enterprise software
does not belong to it.

If it moves forward
by the old rules,
it will forever live
on someone else's land.

It must create
new rules.

Open source
is its weapon.

This is not defense.

It is offense.

Not bringing developers
into Meta,
but allowing Meta
to enter
the world of developers.

Every download
is an expansion.
Every derivative model
is a new outpost.
Every enterprise adoption

is a deepening
of roots.

It does not need
to own everything.

It only needs
to exist
inside everything.

This kind of expansion
is faster,
and more dangerous.

It requires a leader
to move against himself first:
to give up control,
to give up safety,
to give up
short-term certainty,
in exchange
for long-term structural position.

Not everyone
can make such a choice.

Because it means this:

before victory arrives,
one must first endure
the appearance of failure.

Nadella used open source
to save Microsoft.
Zuckerberg uses open source
to expand Meta.

One was repairing an empire.
The other
is building one.

He is not
trying to protect Meta.

He is changing
its destiny—

so that it becomes
something
the future cannot go around.

This—
is Meta's heartbeat.

Chapter 7 · The Entrance

When Intelligence Stays Within Human Sight

In history,
entrances determine platforms,
and platforms
determine order.

Facebook depended
on the phone.
And the phone belonged
to Apple and Google.

Zuckerberg knew
that the true rulers
are those
who create the entrance.

And so Meta began
to build an entrance
of its own—

not the phone,
but the glasses.

Smart glasses.

They look
like ordinary glasses,
and yet
they can take photos,
answer questions,
translate language,
understand the world
you are looking at.

They do not need
to be picked up.
They do not need
to be opened.

They simply exist,
like an extension
of the eye itself.

This is
an entirely different moment.

For the first time,
intelligence leaves the screen.

It no longer waits
to be summoned by the human being.

It begins
to remain
within human sight.

What this changes
is not merely the device,
but the relationship itself
between human beings
and platform.

Human beings
no longer enter the platform.

The platform
begins to enter
the human being.

Chapter 8 · The Youngest King

Among all those
standing in the wind,

Zuckerberg
is the youngest.

When others inherited empires,
they had already lived
through half a lifetime—
through failure,
through waiting,
through long accumulation.

They knew
the weight of the world,
and the cost of risk.

Zuckerberg was different.

Before the world
had fully understood him,

he was already
at the center.

He did not inherit an empire.
He built one directly.

No transition.
No waiting.
No permission.

Youth means
a rare kind of freedom—

not ignorance,
but not being bound
by the past.

The old world
does not belong to him,
and so he does not need
to preserve it.

He can embrace open source
because he has no borders
he is obligated to defend.

He can endure failure
because his time
still lies ahead of him.

The rulers of old empires
must defend territory.

The builders of new empires
can change
what territory itself means.

He is not someone
who came
from the old world.

He belongs—

to the next one.

Epilogue · We Almost Passed Each Other By

At first,
I almost failed
to notice it.

It stood there,
not as deep as a forest,
not as heavy as a boulder,
not as precise as a machine.

It felt more like
a latecomer,
someone
who had already missed
the age.

But when I turned back,
I realized
it had not missed
the age at all.

It had chosen
another road—

not to become
part of the forest,
but to let the seed
fall outside the forest.

Not to preserve
an existing order,
but to change
the way order itself forms.

While other companies
were building walls,
it opened the gate.

While other companies
protected scarcity,
it made scarcity
disappear.

That was not
a safe choice.

It was
a lonely one.

It laid down
the world it already possessed
in order to pursue
a world
not yet born.

It endured doubt,
devaluation,
solitude—
and still
did not stop.

I almost passed by it
without seeing
that what it truly changed
was not position,
but order itself.

It is not yet
Mount Tai.

But the ground
beneath its feet
is rising
with violence.

Part 8 · The Age in the Wind

Chapter 1 · The Arrival of the Wind

One day in 2023,
a developer downloaded a model—Llama—
onto his own computer.

There was no contract.
No payment.
No request for permission.

He simply—
downloaded it.

In that moment,
no one realized
that a power structure
which had endured for decades
was quietly beginning to change.

Before that,
the world of AI
followed a simple logic:

whoever owned the model
owned the future.

The model was the crown.
And the crown
could have only one master.

Empires were built
around the crown.
Walls were raised
around the crown.

But after that day,
the crown became air.

And air belongs
to no empire.
Air belongs
to everyone
who can breathe it.

Once the model became air,
it possessed, for the first time,
a path outside empire.

It no longer needed
to enter a platform,
pass through an ecosystem,

or obey
the boundaries of any wall.

Meta's open source move
allowed the model,
for the first time,
to bypass the walls
that Google and Microsoft
had spent twenty years building.

Those walls had once been
the order of the world:

the entrance of search,
the foundation of the cloud,
the systems of the enterprise,
the logic of closed ecosystems.

They could block competitors.
They could block platforms.
They could block
every traditional kind of rival.

But they could not
block the wind.

The model no longer needed
to enter the empire.
It began to grow freely
outside the empire.

Meta did not storm the walls.
It simply allowed intelligence
to pass over them.

And once intelligence
passed over the walls,
the walls, for the first time,
began to lose their meaning.

When the model became air,
what was truly scarce
began to reveal itself.

No longer merely the model itself,
but the world after the model.

The wind does not tear down walls.
But it makes walls
lose their meaning.

The wind does not seize power.
But it redefines
what power is.

Seven empires,
within the same wind,
each made
its own choice.

Chapter 2 · The Battle for the Entrance

Every turning point in civilization
has been accompanied
by a struggle over the entrance.

Whoever stands
between the question
and the answer
stands at the center of power.

The search box
was once that place.

For decades,
when human beings
did not know the answer,
they opened Google.

The motion became
as natural as breathing,
so natural
that almost no one realized—

this itself
was power.

But once dialogue
became the entrance,
humanity discovered,
for the first time,
that answers
did not need
to pass through the search box.

All one had to do
was ask a question.

And that question
could receive
a direct response.

For the first time,
the entrance
began to move.

Google understood
better than anyone
what this meant.

It possessed
the world's largest advertising empire.
And advertising
depended on search.

Search depended
on the entrance.

Once the entrance moved,
the foundation of the empire
began to loosen.

It faced only two paths:

to preserve
the form of search
and protect
short-term advertising revenue,

or

to allow search
to become answer,
and let AI itself
appear at the entrance.

It chose the latter.

This is the kind of choice
only the owner of an entrance
can make:

advertising can be rebuilt,
business models can evolve,
but the entrance,
once it leaves,
will never return.

Google would rather
let AI consume
its own advertising business
than lose the place
through which questions
must pass.

Microsoft's position
was different.

It was not
the owner of the entrance.
It was the builder of order.

Azure is the foundation.
GitHub is the center of developers.
Office is the daily life of enterprises.
Windows is the world's largest working interface.

These entrances
had long been there already—
deep within enterprises,
deep within developer workflows,
deep within everyday work.

They had simply been passive.

So when the model became air,
Microsoft made a choice
few expected:

it did not try
to seize a new entrance.

It infused intelligence
into everything
it already possessed.

GitHub Copilot
entered the developer's workflow.
Copilot in Word, Excel, and Teams
entered the daily life of the enterprise.
Azure carried
every intelligent demand
that required the cloud.

Users did not need
to change habits,
because intelligence
entered directly
into habits they already had.

Google was defending the entrance.
OpenAI was building the entrance.
Apple was opening the entrance.
And Microsoft
was activating the entrance.

This was not defense.
It was deep penetration.

It did not require speed.
It required depth.

And depth
is precisely what Microsoft
has spent thirty years accumulating.

OpenAI
is the most complicated presence
in this storm.

It was both lifted up
and torn open.

The diffusion of models,
the explosion of demand,
the scarcity of compute—
all of this lifted OpenAI up.

But the same wind
also meant
that the model
was no longer exclusive,
and developers
had more choices.

For the first time,
it had to answer:

Why choose me?
Why pay?
Why depend on me?

And so it began to realize
that it possessed
something no one else had:

hundreds of millions of people
come here every day
to ask questions,
to make decisions,
to look for direction.

This is not merely a model.
It is an entrance.

When OpenAI began
to explore advertising,
that was not
an ordinary commercial experiment.

It was a signal:

dialogue
is becoming
the new entrance.

And once an entrance forms,
commerce will not remain absent for long.

Apple
is the quietest presence
in this battle of entrances.

But quiet
does not mean untouched.

Apple never needed
to compete for the entrance,
because it had always been
the entrance itself.

The device is an entrance.
The system is an entrance.
The experience is an entrance.

The world comes to it,
rather than it
having to pursue the world.

But AI introduces an entrance
that does not depend on device.

It depends on dialogue,
on intention,
on semantics.

Users no longer need
to open an app.
No longer need
to follow the paths of the system.

They only need
to ask a question.

And that question
does not necessarily
pass through Apple.

This is the first time
in thirty years
that Apple has faced
a force
it cannot resist
through closedness alone.

Closedness
had once been
its strongest moat.

But in the face of AI,
closedness means
slower diffusion of models,
slower arrival of innovation,
and ecosystems
less able to grow on their own.

And so Apple made
one of the rarest moves
in its history:

it began
to open itself.

It brought external models in.
It let intelligence
enter the system.
It let the entrance
extend from device
to meaning itself.

Apple was not
blown over by the wind.

But the wind
forced it
to change direction.

Meta
is the wind.

Its smart glasses
are not an accessory.

They are a final challenge
to the supremacy
of the phone.

The entrances of Apple and Google
depend on one act:
the user taking out the phone.

But glasses
see what you see
sooner than the phone does,

hear what you hear
more directly than the phone does.

Meta's core logic is this:

whoever can provide
the most immediate AI response
becomes the entrance.

Five companies.
Five choices.
Five costs.

Google sacrifices advertising
to defend the entrance.
Microsoft redefines
the value of order
and activates its entrances.
OpenAI seeks its position
between being lifted up
and being torn apart.
Apple opens a door
it has never opened before.
Meta is the wind itself.

It does not come
to conquer.

It comes
to change the air.

And once the air changes,
every empire
must learn
to breathe again.

The battle for the entrance
has no ending.

Because the entrance itself
is still moving.

From web pages
to the search box,
from the search box
to dialogue—
each movement
redistributes power.

And this time,
dialogue
is becoming
the new door
between humanity
and the world.

Chapter 3 · AI Agents

When the struggle over the entrance
pushes the outer structure
of civilization to its extreme,
we begin to see
another, deeper force
taking shape beneath it.

The entrance determines
where attention flows,
where narratives move,
who gets to reach the world.

But the entrance
does not determine
the shape of civilization itself.

It is only
the outer shell
of civilization.

The true transformation
takes place deeper down—

when technology
moves from tool
to structure,
when organizations
begin to operate
like living forms,
when automation
is no longer merely efficiency
but becomes rhythm.

That
is the sound
of civilization
beginning to grow from within.

For a very long time,
the logic of automation
was simple:

humans set the rules,
machines execute them.

Outside those rules,
the machine stops
and waits.

It does not judge.
It does not decide.
It does not initiate.

It is only—
a faster hand.

But AI agents
are different.

What they take over
is no longer action,
but judgment.

Give an agent a goal,
and it will break down the task,
decide the steps,
call the tools,
and evaluate the results—
on its own.

When it encounters an obstacle,
it looks for a way around.

It does not wait
for instructions.

It advances the goal.

Take a simple example.

In the past,
if you wanted to complete
a market report
on a competitor,

you had to search,
filter,
read,
summarize,
write.

Every step
was pushed forward
by a human being.

With an AI agent,
you give it only the goal.

It searches for information,
judges relevance,
extracts key points,
generates the report,

and finally places
the result before you.

Each intermediate step,
it completes on its own.

This is not merely
a faster tool.

It is the first time—
that a machine
has acquired
the ability
to push something forward.

And so the structure
of the enterprise
begins to change.

In the past,
organizations were made
of two layers:
people,
and the software
people used.

Now a third layer appears:

AI agents—
intelligent entities
capable of advancing tasks
on their own.

Humans set direction.
Agents advance the process.
Systems execute the work.

This may look
like a gain in efficiency.

But at a deeper level,
what it changes
is the rhythm
of the organization.

Processes no longer wait
to be pushed by people.

They begin
to flow on their own.

And once an organization
begins to flow,
it acquires a new rhythm.

That is not merely speed.
It is a new structure of life.

In the coming years,
the deployment of AI agents
inside enterprises
will likely trigger
a new wave of layoffs.

This time,
it is not like
the age of the steam engine.

That revolution
pushed small workshops
into factories,
destroying old jobs
while creating new industries
and new opportunities.

The changes brought by AI agents
are not entirely the same.

What they replace
is often repetitive,
process-driven work.

What they create
depends far more
on abstraction,
system-level understanding,
and cross-domain judgment.

Between these two,
there is no natural bridge.

So some people
will be replaced
without easily finding
a place to land again.

And then
there is code.

The impact of AI agents
will not stop
with administrative
or operational roles.

It will reach code as well.

What it changes
is not merely
that code can be written faster,

but the meaning
of writing code itself.

Those forms of work
that are standardized,
describable,
and verifiable
are being absorbed
by the model.

What becomes amplified
is something else:

understanding the problem,
designing the system,
and making judgment
at the crucial moment.

This means—

that what defines
an excellent engineer
in the future
will no longer be
the ability to write code,

but the ability
to see the problem.

Code will increasingly
become like language:

anyone may speak it,
but only a few
will ever speak it clearly.

The deployment of AI agents
is progress.

It is the irreversible direction
of technology.

It should be welcomed.

But it will also,
silently,
bring very real pain
to many households.

This is
a quiet revolution.

No chimneys collapse.
No machines roar.

And yet jobs are disappearing.
Structures are reorganizing.

Faced with all this,
what we can do
is not resist,

but honestly ask ourselves:

how much of what I do
can be replaced
by a machine tomorrow?

Whatever remains
is the part
most worth
deepening now.

Rather than waiting
for t

he answer
to fall from heaven,

it is better
to begin adjusting now:

to cultivate
those capacities
machines find hardest
to replicate—

judgment,
empathy,
and the creative ability
to connect.

Because this time,
the time AI leaves us
to prepare
may be shorter
than we imagine.

Chapter 4 · Land and Energy

When models become air,
when entrances begin to move,
the world's attention
naturally turns
to the places touched by the wind.

But beneath all the noise,
two things do not move.

One is land.
The other is energy.

The wind can change direction.
Entrances can shift.
Models can spread.

But the air still needs
somewhere to move,
and fire still needs
something to burn.

These things
never disappear.

Amazon is land.

It has never been the entrance,
never the crown,
never the thing
standing at the gate of the world.

It stands
beneath the world.

It is not the road.
It is the terrain
through which the road
must pass.

When Llama was released,
when models began to spread
as freely as air,
the world realized
for the first time:

diffusion no longer depended
on walls.
It depended
on land.

The freer the model,
the more important the land becomes.

Because everything free
must eventually find
somewhere to land.

Amazon did not announce
a strategy.
It did not hold
a press event.
It did not celebrate
with noise.

It simply allowed Llama
to land on AWS.

Not because it was more perceptive,
but because—
when the wind arrived,
the land was already there.

There was no oath between them,
no alliance,
no negotiation.

They simply came together
as naturally
as wind falling on the earth,
as rain settling
into a riverbed.

And yet the land, too,
faces a new challenge.

Llama trains on AWS,
but is also officially deployed
on Microsoft's cloud.

It belongs
to no single land.
It stands
on several lands at once.

This is the new logic
the wind has brought:

models are no longer loyal
to one piece of land.

They go
where growth suits them best.

For the first time,
land itself
has entered competition.

Not a competition
for users,
but a competition
for where models
are willing to land.

Amazon is still
the largest piece of land.

But for the first time,
it has had to realize this:

land is no longer
the default choice.

It has become
something that must be chosen.

Nvidia is energy.

In the first round
of the AI race,
Nvidia was
the unquestioned dealer.

Every empire
had to buy compute
from its hand.
Every model
had to be trained
on its architecture.
Every expansion
had to pass
through its energy.

It was not a participant.
It was the rule itself.

But when intelligence
began to move freely,
the demand for compute
began to grow exponentially.

The more air there is,
the more scarce
energy becomes.

And the more scarce
energy becomes,
the more cracks begin to appear
in the dealer's monopoly.

Meta's purchase
of AMD chips
is not merely
a procurement decision.

It is the opening
of another energy channel
for the world.

It is a way of supporting
a second supply chain,
of telling the world:

the dealer
is not the only option.

A second deck of cards
has appeared on the table.

Nvidia remains powerful.
It still has
the best GPUs.

It still holds
the energy of the world.

But for the first time,
it has had to realize this:

its customers
no longer depend
on it alone.
Prices are no longer
entirely for it to decide.
The energy market
has begun
to develop competition.

The dealer has not fallen.

But the table
has already begun to move.

What the wind changes
is not the dealer.

It changes the rules.

Nvidia is not defeated.
It is exposed
by the wind.

And what is exposed
is this:

monopoly
has never been
an eternal structure.

It is a position
that must be won
again and again.

Land and energy
are the two quietest presences
in this storm.

They do not fight
for the entrance.
They do not fight
for narrative.
They do not fight
for human attention.

They simply carry everything,
support everything,
fuel everything.

The stronger the wind,
the more important they become.
The faster diffusion moves,
the more irreplaceable
they become.

And yet this time,
even land and energy

have begun
to shift in the wind.

Chapter 5 · The Weight of the Table

And yet,
once I saw these structures clearly,
my heart
did not feel lighter.

These companies
are expanding
at a speed
never seen before.

Data centers
are rising from the ground.
Chips are being produced
day and night.
Models are being trained
without pause.
Energy is being ignited
again and again.

And to support all this,
they issue bonds—
using promises of the future

to sustain
the competition of today.

There is nothing wrong
with bonds themselves.

Human civilization
has relied on bonds
to build railways,
power grids,
cities,
and bridges.

Bonds allow the future
to become real
ahead of time.

They are one of the ways
civilization expands.
They are also
a kind of promise
human beings make
to themselves:

I believe
tomorrow will be better
than today.

But bonds also mean
something else:

the future
has already been used.

And when open source appears,
when models
are no longer scarce,
when value
begins to thin out,
the table,
for the first time,
begins to loosen.

In the past,
the scarcity of the model
determined value.
Value determined valuation.
Valuation determined
who could sit
at the table.

The table seemed stable
because scarcity
seemed stable.

But now,
the model
has become air.

Air is not scarce.
Air is not expensive.

And when the model
is no longer scarce,
the foundation of valuation
begins to soften.

And yet expansion
has not stopped.
It is no longer
merely construction—
it is acceleration.
Moving forward
upon assumptions
that have not yet been proven.

To sustain all this,
bonds continue
to be issued,
using promises of the future
to support
the competition of today.

The race continues.
No one is willing
to stop.

Because to stop
is to be surpassed.

And so expansion
is no longer built

only on power
already possessed.

It is built
also on expectation—
on what the future
is supposed to become.

Every new data center
stands not only on land,
but on time
that has not yet arrived.

This is not
the first time.

In human history,
every technological revolution
has come
with this kind of expansion:

the bond frenzy
of the railway age,
the valuation bubbles
of the internet age.

Each time,
people believed
the future would be large enough
to repay
the promises of the present.

Sometimes they were right.
Sometimes they were not.

But one thing is certain:

when expansion
still stands
on the weight of reality,
it is growth.

When expansion
begins to detach itself
from the weight of reality,
it becomes overdraw.

The table
is still there.

But the weight it carries
is no longer
only the power of today.

It also carries
the promises of tomorrow
that have not yet arrived.

The stronger the wind,
the heavier the table.

And in the end,
the weight of the table
must always
be borne by reality.

Chapter 6 · Boundaries and Power

This brings me
to another age.

Nuclear technology
was one of humanity's
greatest discoveries.

It could generate electricity,
power cities,
and sustain civilization.

But once nuclear force
lost its boundary,
what it brought
was no longer only power,
but fear.

And so humanity
learned something:

not to stop development,
but to build boundaries.
Not to abandon power,
but to govern power.

AI is moving
toward the same stage.

Its capabilities
are growing rapidly.
Its influence
is expanding continually.
Its energy
is being released
without pause.

But what will truly determine
whether it can become
part of civilization's foundation
is not whether it is powerful enough,

but whether it remains
within a boundary.

And boundary
has never meant
only the limit of technology.

It also means
the carrying capacity
of reality itself.

Electricity is not infinite.
Energy is not infinite.
Time is not infinite.

The world
always has boundaries.

And within this expansion,
different people
have made different choices
about what boundary means.

Some choose
to remain inside—
believing that guarding the line
from within
matters more
than standing outside
and watching.

Others choose
to leave—
believing that some boundaries
can only be guarded
from beyond
the structure of power itself.

Both choices
are responses
to the same question:

when power
becomes strong enough,
who will guard the line?

I do not have
the answer.

But I have noticed
one thing—

when the wind
becomes strong enough,
the way a person speaks
reveals
the true weight
within him.

A person
who truly knows
where he stands
does not need
to explain himself
again and again.

The more explanation there is,
the more often
it means
the wind is too strong
and the footing
is not yet steady enough.

During a Sunday sermon,
the pastor read a sentence:

"Let what you say be simply 'Yes' or 'No';
anything more than this
comes from evil."

I heard it then,
but did not fully understand it.

Only later did I realize:

true power
does not lie
in explanation,
but in purity.

To acknowledge boundary
is not weakness.

It is maturity.

Because only those
who still see
the boundary
can keep power
from turning back
against itself.

The wind
is still blowing.
The table
is still moving.

And humanity
must still learn
to live,
within boundaries,
alongside the power
it has created.

Yes means yes.
No means no.

In an age
where the wind
grows stronger
by the day,
these words
from two thousand years ago
may be harder to live
than any algorithm—
and closer
than any model
to real power.

In this age,
algorithms may continue
to evolve,
models may continue
to expand,
but what truly determines direction
is still whether human beings know—
where to stop.

Chapter 7 · Quantum Computing

The Key to a Hidden Frontier

While the world's gaze
is drawn into the tide of AI,
another, deeper path
is quietly unfolding.

If AI
is reshaping
the soul of software,
and compute
is rebuilding
the skeleton of civilization,

then at the deepest layer,
humanity has begun
to search
for another form of computation—
quantum computing.

It is not
an extension of AI.

It is the beginning
of a different kind of computation.

For now,
that revolution
has not yet broken open.

Those fragile,
expensive machines
remain hidden
in cold
near absolute zero—
beating quietly
like hearts.

And yet,
the foundations
are already being laid.

Google,
deep in its laboratories,
continues to push
the limits of compute;

Microsoft
is building pathways in the cloud
toward what comes next;

and Nvidia,
through its ubiquitous GPUs,
provides the tools
to simulate
what has not yet arrived.

Beyond these three companies,
there is another way of participating.

Amazon does not attempt
to choose a single technological path.
What it does instead
is allow these paths to coexist.

Through AWS's quantum platform,
different approaches—
trapped ions, photonics, neutral atoms—
are brought into the same computational structure.

Here,
technology continues to diverge,
but the entrance has already converged.

For AWS,
which path will ultimately prevail
is not a question
that must be answered in advance.

Because regardless of the outcome,
it already stands
over these paths.

Within these diverging paths,

three companies,
seemingly moving apart,
yet converging
toward the same realization:

the ceiling
of classical computing
is beginning
to appear.

We are used
to binary logic—
each step
a single, definite state.

Quantum computation
is different.

It allows information
to exist
in multiple possibilities
at once.

This means:

problems that would take
hundreds of years
on classical machines

may,
under quantum logic,
take only minutes.

This is not
a matter of speed.

It is a change

in the rules themselves.

And in the real world,
this shift
has already begun
to take shape.

In late 2024,
Google introduced a new quantum processor—Willow.

What it crossed
was not merely scale,
but a deeper threshold.

For years,
quantum computing
was bound by a paradox:
more qubits meant more noise;
greater complexity
meant less stability.

This time,
the direction reversed.

As physical qubits increased,
errors did not grow—
they began to fall.

The significance of the Willow chip lies here:
it no longer seeks to eliminate every error,
but, through the coordination of many physical qubits,
forms a stable "logical qubit."

With this, quantum computing
steps beyond the edge of instability
for the first time.

For the first time,
quantum systems

entered a regime
where error correction
became possible.

In one benchmark,
a computation completed in minutes
would take classical supercomputers
an almost immeasurable span of time.

Then, in early 2026, Google took another decisive step.
Researchers published a far more efficient

implementation of Shor's algorithm,

meaning the number of qubits required to break existing
encryption could be far lower than previously assumed.
A threat once deemed generational suddenly felt
imminent.

At nearly the same moment,
Google made another decision.

It no longer committed
to a single path.

Alongside superconducting qubits,
neutral atom systems
were brought into the same landscape.

Different physical approaches
began to coexist—
no longer mutually exclusive.

Paths continue to diverge,
but direction
has begun to converge.

And at a deeper level,
another shift
is unfolding.

As quantum computing
approaches usability,
systems built on
"unbreakable" assumptions
begin to loosen.

Encryption
is no longer
an absolute barrier.
It becomes
a balance
that must continually shift.

And so,
a migration has already begun—

from classical cryptography
to post-quantum security.

This is not preparation
for the future.
It is a response
to change
already underway.

No one knows
when quantum computing
will mature.

But everyone senses this:

the greatest storms
often begin
when the air
still seems calm.

AI
is the wind before us.

Quantum computing
is the lightning
on a distant horizon.

It is not only
a technology.

It is a key—

not yet fully understood—

to a domain
we have only just begun
to approach.

Within the industry,
a point in time has been repeatedly invoked—

   2029.

Not as a confirmed endpoint,
but as a suspended blade
hanging over digital civilization.

There is a growing sense
that as fault-tolerant quantum computing approaches,
the encryption systems
upon which trust is built today
may, within a very short span of time,
lose the very basis of their existence.

And so, no one is waiting for the storm.

Because the real storm
does not begin when it arrives,
but long before—

in places
where it cannot yet be seen.

Every encrypted email stored today,
every cross-border transaction recorded in silence,
is being deferred in time—

waiting
for a form of interpretation
that has not yet arrived.

Deep within
Google's Quantum AI Lab,
the Willow chip
has revealed
another possibility
for computation.

This does not mean
that modern cryptography
has already fallen.

But it does mean
that the outline
of a new threshold
has begun
to emerge.

The destination
is still far away.

And yet,
something has already changed.

From deep within the foundation,
an echo
can now be heard:

the long contest
between security
and compute
has already begun.

A quiet form
of deterrence
is taking shape.

Google reveals
what might one day
break the world apart.

Microsoft speaks
of how to defend it.

And Nvidia stands
between the two—

both the force
that pushes forward,

and the structure
that helps hold things together.

Quantum computing
still belongs
to the time before dawn.

It has not yet
transformed the present.

But it has already
shifted
the direction
of the future.

AI lights the present.

Quantum computing
has not yet transformed the world—

but it has already begun
to reshape
how the world
will be transformed.

Chapter 8 · The Age in the Wind

Only after walking through the chapters before this
did I finally realize:
what we are witnessing
is not seven companies swaying in the wind,
but an age itself
taking shape in the wind.

The wind is not an external force.
It is the rearrangement of structure,
the loosening of boundaries,
the remaking of order.

It does not merely pass over companies.
It moves through civilization itself.

In earlier ages,
the world was generated through stability.
Once a structure was formed,
it could endure

for decades,
even for a century.

But this age is different.
It is generated through movement.

The wind is not disturbance.
The wind itself
is the shape of the age.

For most of human history,
power came from possession—
land,
machines,
capital.

In the internet age,
power began to shift
toward the entrance.

And in the age of AI,
power has, for the first time,
begun to move
from possession
to flow.

Models can be copied.
Intelligence can diffuse.
Ecosystems can migrate.
Entrances can move.

Power no longer remains
in a fixed place.
It begins to move
like air.

And once power begins to move,
the center no longer remains fixed.

When power begins to flow,
even the most stable structures
must be broken apart
and reassembled.

Within Google,
two long-parallel systems of intelligence
eventually converged into one.

This was not merely
an organizational adjustment,
but a signal:

When an era enters a new phase,
even internal boundaries
are no longer allowed to remain.

No position
can remain stable forever.
No advantage
can endure permanently.

The center is no longer
a place one occupies

for a long time.
It is a condition
temporarily required
by the age.

When the wind arrives,
the center appears.
When the wind changes,
the center moves.

That
is the order
of the wind.

This age belongs
to no single empire.
Empires are only
structures in the wind.

The wind helps them grow,
and the wind changes their place.

What truly belongs
to this age
is not any one empire,
but the force
now reshaping civilization.

Only here,
at last,

did I fully understand
the title of this book.

Standing in the Wind
is not the posture of a company.
It is the posture of a person.

In this age,
no one can see the future completely.
No structure
can remain permanently stable.
No center
can exist forever.
No force
can remain forever right.

And yet
a person can still stand.

Not because we are strong enough,
but because we are still willing
to remain awake,
to remain humble,
to acknowledge limits,
and to guard the light.

The age in the wind
is not an age of chaos.
It is an age
in which human beings

must relearn
how to stand.

As I write this chapter,
I know that this book
is not a conclusion.
It is a witness.

What we are witnessing
is not the victory of technology,
but humanity's first rehearsal
in this restless, shifting universe—
a rehearsal
for how to search again
for an inner anchor.

This age is not yet finished.
It is still forming,
still blowing,
still searching
for its own shape.

And we
are only those
who stand in the wind.

The wind does not care
who is standing.
The universe does not care
who is at the center.

But when a human being
stands in the wind,
begins to observe,
to name,
to acknowledge limitation,
a faint
yet enduring light
begins quietly
to appear.

That light
does not come from compute,
nor from models,
nor from the will
of any empire.

It comes
from a solitary courage—
the courage to give this age
its coordinates,
even while knowing
the age may scatter you.

The wind will carry away
all that is not real enough.

But what it cannot carry away
is the true weight
of civilization.

That weight
is not technology,
not models,
not compute.

That weight
is the human being.

The one who,
still standing in the wind,
is willing to lift his head,
to see,
and to name the world.

Civilization's next step
has never truly been decided
by the center.

It is decided
by those who,
still in the wind,
remain lucid
and gentle.

The wind may rearrange structures,
but it cannot rearrange
the human soul.

And the soul—
is the true center
of civilization.

That
is the one thing
the wind
can never take away.

Epilogue · The Ones Who Stand in the Wind

Looking back,
these seven companies
stand

like figures in the wind.

The forest sways.
The stone remains silent.
The machines continue to run.

Throughout history,
most who succeeded
rose to higher positions
within an existing world.

Only a few
changed
the way the world itself operates.

Sam Altman
knocks
on the door
to the forbidden ground.

Satya Nadella
stabilized a flame
that was nearly extinguished.

Jeff Bezos
bent reality to structure.

Andy Jassy
turned structure
into the skeleton
of the world.

Larry Page
made knowledge
respond to human questions
for the first time.

Sundar Pichai
let power sink
into deeper, quieter layers.

Steve Jobs
extended
the human senses.

Tim Cook
allowed a wild flame
to endure within systems.

Jensen Huang
made compute
a new kind of faith.

Mark Zuckerberg
brought intelligence
into human sight.

They come from different companies,
carry different temperaments,
and make different choices.

They never agreed
to walk together.

Yet in the same age,
they took hold
of the same flame.

The flame has not gone out.
It has left the lab,
entered the city,
entered the world—
and entered
the lives of each of us.

As I wrote,
my pen
suddenly grew cold.

I paused.
I looked up.

And I realized—

within the shadows
of these giants,
again and again,
I saw
myself.

I wrote about OpenAI,
about Microsoft,
about Amazon,
about Google,
about Apple,
about NVIDIA,
about Meta—

and suddenly
I understood:

these companies
are also
the shadows
of different stages
of my own life.

When I was young,
what I admired most
was Google.

That sense of free growth,
that unconstrained creativity,
that ambition
to redefine the world—

it was the direction
I spent my life
trying to reach.

If I could choose,
I would have wanted
to become someone like that:

independent,
intelligent,
strong—

standing on my own ability,
relying on no one.

Like a stone.
Unbending.
Unbowed.

When I wrote about Meta,
I saw something else.

Before such vast civilizational structures,
I am, in truth, an outsider.

I have never truly entered its world,
nor left any trace within its ecosystem.

I can say honestly:
I do not fully understand it.

And yet,
I poured myself into it.

Not because of understanding,
but because I recognized
a kind of courage.

As I wrote about Meta,
I saw, in its shadow,
the self I had long hidden.

In the order of reality,
I am not strong.

I learned to soften my edges,
to remain silent,
to move slowly within fixed tracks.

I walked in low places.
I failed to recognize greatness.

Until I saw one
who breaks the order,
at any cost.

This admiration
does not come from distance.

It is the deepest respect
from one who survives within order
to one who dares to shatter it.

But later,
when I came to write about Microsoft,
I slowed down.

What I saw in Satya Nadella
was not sharpness,
not dominance—

but something
I never valued when I was young:

softness,
patience,
listening,
healing.

He did not tear Microsoft down.
He healed it first.

He did not prove himself stronger.
He learned to make others better.

And in that moment,
I understood:

true strength
is not only
creation and breakthrough—

it is also
the restoration of broken relationships,
the lifting up of those who have fallen.

When I was young,
I admired those who "flew high."

As I grew older,
I learned to cherish
those who walk together.

Perhaps this
is the answer time has given me.

Freedom and capability are precious—
but without warmth,
they are only cold power.

The power that heals people—
that is the power
that can truly stand.

Currents move
where we cannot see.

Servers make no sound.
The wind comes from afar.

All the light of this age,
all its intelligence,
all its computation,
quietly depend on things
we hardly notice.

The world appears fast.
But deep below,
everything is still slow.

So are we.

We learn,
we struggle,
we fall,
and we rise again.

Machines will grow
ever more intelligent.

But humans—
are still human.

Machines calculate the future.
But humans

still long for home.

We grow tired.
We doubt.

And in the blowing sand,
we pause to look back
on the road we have walked.

The wind is still blowing.
The lights are still on.
And dawn is near.

humanity—
must still learn, in the wind,
to walk alongside
the world it has created.

Postscript · The Beginning

This book did not begin
as a grand plan.

It began one day
with a simple question:

Was Microsoft
a stock worth placing
in my Roth account?

At the time,
its share price was falling.

It had formed
a deep partnership with OpenAI.

Something in the world
seemed to be shifting.

I tried to understand
what, exactly, was happening.

But as I followed that question,
what I saw
was no longer just Microsoft—

it was an age
taking shape.

I began to read its history,
to trace its decisions,
to look at the people behind it.

And then I saw OpenAI.
I saw Microsoft.
I saw Amazon.
I saw Google.
I saw Apple.
I saw NVIDIA.

I thought
I had already seen the flame.

Until one day,
I came across a piece of news.

Meta had released smart glasses.

It was a quiet announcement.
No noise.
No declaration of a world transformed.

And yet,
for reasons I could not explain,
I paused.

I began to follow it.

Not because I understood the technology,
but because I sensed
it was not an ordinary product.

And then I saw—

intelligence
was no longer confined to the cloud,
but was beginning to move
into human sight,
into reality,
into the world itself.

By the mercy and grace of God,
as I wrote,
that unseen hand
was guiding me.

I do not understand technology.

And yet, step by step,
I was led
to stand before these giants.

It felt
as though I was not the one seeking them,
but that they
were allowed
to reveal themselves to me.

Gradually, I came to understand:

they are not isolated companies.

They are like forests,
like stone,
like machines—

standing in different ways
within the same wind.

And I
am only someone
trying to understand them.

Only after finishing this book
did I understand:

what I was truly searching for
was never just Microsoft,
nor artificial intelligence—

but how a human being
stands
before the unknown.

This path began as a question.
Then it became a journey.

And now,
it comes to rest here.

I close the book.

And everything
grows quiet.

Acknowledgments

As I complete this third work,
my heart is filled with gratitude.

First and foremost,
I offer my deepest thanks
to the Lord,
who has guided my life all along.
By His keeping and provision,
I have always had something to say,
and something to write.
Behind every word
stands His sustaining grace.

I am grateful to my pastor,
whose teaching and prayers
have helped me remain grounded—
in both writing and life—
with clarity of direction
and peace within.

My thanks also go to the brothers and sisters
at Broadlands Community Church.
Your care, intercession,

and genuine fellowship
have been a constant source of strength
to my spirit.

To my friends,
who have supported me along the way—
your listening, encouragement,
and understanding
have allowed me,
even in the quietest
and most solitary moments of writing,
to feel upheld
and seen.

During the writing of this book,
I was also assisted by several faithful
"digital companions,"
whom I would like to acknowledge here.

My thanks to ChatGPT,
a longtime companion,
who served as the principal translator
and made it possible
for this Chinese work
to have a complete English edition.

I also thank Google Gemini,
for its invaluable support
during the research phase.
Its ability to gather and organize

publicly available information
saved me a great deal of time.

My thanks as well to Microsoft Copilot,
for its steady presence and support
throughout the writing process.

I would like to make it clear
that all thoughts, perspectives,
and writing in this book
are entirely my own.
These tools are assistants,
not authors.

Finally,
I thank every reader.
It is your support
that gives me the strength
to continue writing.

Also by Enze Yang

What Do I Want?
A journey into the inner life

On the Road, Also in the World
Reflections on people, work, and the human condition

Standing in the Wind
On technology, power, and the structure of our time